AF593595

Agroecosystems Analysis

Agroecosystems Analysis

Co-Editors
Diane Rickerl
Charles Francis

Editorial Committee
Robert Aiken
C. Wayne Honeycutt
Frederick Magdoff
Ricardo Salvador
Charles Francis
Diane Rickerl

Managing Editor: Lisa Al-Amoodi

Editor-in-Chief ASA Publications: Kenneth A. Barbarick

Editor-in-Chief CSSA Publications: Craig A. Roberts

Editor-in-Chief SSSA Publications: Warren A. Dick

Number 43 in the series
AGRONOMY

American Society of Agronomy, Inc.
Crop Science Society of America, Inc.
Soil Science Society of America, Inc.
Publishers
Madison, Wisconsin, USA
2004

American Society of Agronomy, Inc.
Crop Science Society of America, Inc.
Soil Science Society of America, Inc.
677 South Segoe Road, Madison, WI 53711-1086 USA

ISBN: 0-89118-153-9

Library of Congress Control Number: 2003116111

Printed in the United States of America.

CONTENTS

FOREWORD

Agroecology is a scientific discipline that seeks to provide an objective, ecologically based assessment of the structure, function, multidimensionality, and spatial scale of food systems. It is a complex science, one that links the ecological, economic, social, ethical, and legal aspects of food production. All spatial scales are considered, from farm field to global, and systems approaches are emphasized. This holistic approach to the analysis of agricultural systems is increasingly necessary, given the growing complexity of problems we face in producing food for an expanding world population. Many contemporary agricultural issues, such as agricultural impacts on soil, water, and air quality; conflicts related to water conservation and the availability of water for agricultural use; global warming and carbon sequestration; the advent of genetically modified plants and animals; the political and ethical concerns related to the globalization of the agricultural economy; and the changing social fabric of rural agriculture can only be addressed by a broader, systems-based approach to research, education, and outreach.

This ASA–CSSA–SSSA monograph provides an excellent overview of the current state of the science of agroecology. Leaders in this field give detailed analyses of key topics such as multidimensional thinking, multifunctional economic analysis, whole-farm planning, agricultural conservation at the landscape scale, agroecosystem functions that benefit society, and ecological morality. A critical analysis of our current approach to designing agricultural research, teaching, and funding programs is provided, followed by an alternative vision of how we should redesign these programs in more holistic, sustainable manners. Research and education priorities, suggested across a spectrum of spatial scales for the full breadth of key topics in agroecology, will be valuable guidance for farmers, researchers, teachers, students, and policy-makers.

J. THOMAS SIMS
President
Soil Science Society of America

LOWELL E. MOSER
President
American Society of Agronomy

KENNETH J. MOORE
President
Crop Science Society of America

PREFACE

Scientists and educators have long struggled with the complexity of learning about agricultural and food systems. *Agroecosystems Analysis* provides practical and scientifically based guidelines as well as specific indicators of agroecosystem structure and function to help students and professionals unravel this complexity. We approach the topics across a hierarchy of spatial scale from the microbiological to the landscape and regional levels. This perspective is important to promote better understanding of the flow of agroecosystem functions from the physical structuring of the field, the farm, the landscape, and the region.

Most agricultural scientists and many of our courses in current curricula focus on the production aspects of agriculture. The goals of increased crop yields and animal production are inherent and assumed in the design of cropping, grazing, and crop–animal systems. Yet we know that sustainability of agriculture and food systems will depend on more than production. For this reason, the indicators used to assess systems in the following chapters reflect biological, ecological, economic, and social dimensions of agroecosystems.

The science of agroecology has developed as a framework within which we can objectively measure and evaluate food and fiber systems. The editors and authors of *Agroecosystems Analysis* intend the monograph to serve as both a professional research reference and a textbook for advanced undergraduates or graduate students. Each chapter contains study questions. The monograph concludes with a glossary and subject index.

In order to integrate and build upon knowledge gained from different disciplines and perspectives, the chapter authors exchanged information during preparation, writing, and editing their contributions. This process helped us to integrate the material and add multiple dimensions to the complex systems we all strive to understand.

As editors, we have provided a framework for chapter author creativity. We have also attempted in the first and last chapters to take the conversation outside the proverbial box in order to stimulate the reader to go beyond what is written here. We anticipate that students will develop new concepts, make new connections, and thus advance the study of agroecosystems beyond the horizon that has been provided by the authors in this humble attempt to take a snapshot of systems today. We look forward to participate with you in that search.

The editors wish to acknowledge the contributions of the editorial committee members appointed by the American Society of Agronomy, Crop Science Society of America, and Soil Science Society of America—Ricardo Salvador, Wayne Honeycutt, Rob Aiken, and Fred Magdoff. They met with the editors and some authors on several occasions, and participated in formal review of book chapters. We gratefully acknowledge the contributions of Patricia Wieland to the preparation of the final manuscript.

Co-Editors

DIANE RICKERL
South Dakota State University
Brookings, South Dakota

CHARLES FRANCIS
University of Nebraska
Lincoln, Nebraska

CONTRIBUTORS

Miguel A. Altieri Division of Insect Biology, University of California at Berkeley, 201 Wellman Hall 3112, Berkeley, CA 94720

Robert M. Caldwell School of Natural Resources, University of Nebraska, 306 Keim Hall, P.O. Box 830910, Lincoln, NE 68583-0910

Thomas L. Dobbs 109 Scobey Hall, Box 504, Economics Department, South Dakota State University, Brookings, SD 57007-0895

Cornelia Butler Flora North Central Regional Center for Rural Development, 107 Curtiss Hall, Iowa State University, Ames, IA 50011-1050

Charles Francis 102B KCR, Department of Agronomy and Horticulture, University of Nebraska-Lincoln, P.O. Box 830915, Lincoln, NE 68583-0915

Stephen R. Gliessman Program in Community and Agroecology, Department of Environmental Studies, University of California, Santa Cruz, CA 95064

Juha Helenius Department of Plant Production, University of Helsinki, P.O. Box 27, Viiki FIN-0014, Finland

Rhonda R. Janke Sustainable Cropping Systems, Department of Horticulture, Forestry, and Recreation Resources, 2021 Throckmorton Hall, Kansas State University, Manhattan, KS 66506

Frederick Kirschenmann Leopold Center for Sustainable Agriculture, 209 Curtiss Hall, Iowa State University, Ames, IA 50011-1050

Geir Lieblein Department of Plant and Environmental Sciences, NLH, Agricultural University of Norway, P.O. Box 5022, N-1432 Ås, Norway

Clara I. Nicholls Division of Insect Biology, University of California at Berkeley, 201 Wellman Hall 3112, Berkeley, CA 94720

Diane Rickerl NPB 247B Box 2140C, South Dakota State University, Brookings, SD 57007

Lennart Salomonsson Centre for Sustainable Agriculture, P.O. Box 7005, Swedish University of Agricultural Sciences, S-75007 Uppsala, Sweden

Thomas E. Schumacher NPB 247 A, Box 2140C, South Dakota State University, Brookings, SD 57007

Conversion Factors for SI and non-SI Units

Conversion Factors for SI and non-SI Units

To convert Column 1 into Column 2, multiply by	Column 1 SI Unit	Column 2 non-SI Units	To convert Column 2 into Column 1, multiply by
	Length		
0.621	kilometer, km (10^3 m)	mile, mi	1.609
1.094	meter, m	yard, yd	0.914
3.28	meter, m	foot, ft	0.304
1.0	micrometer, μm (10^{-6} m)	micron, μ	1.0
3.94×10^{-2}	millimeter, mm (10^{-3} m)	inch, in	25.4
10	nanometer, nm (10^{-9} m)	Angstrom, Å	0.1
	Area		
2.47	hectare, ha	acre	0.405
247	square kilometer, km^2 (10^3 m)^2	acre	4.05×10^{-3}
0.386	square kilometer, km^2 (10^3 m)^2	square mile, mi^2	2.590
2.47×10^{-4}	square meter, m^2	acre	4.05×10^3
10.76	square meter, m^2	square foot, ft^2	9.29×10^{-2}
1.55×10^{-3}	square millimeter, mm^2 (10^{-3} m)^2	square inch, in^2	645
	Volume		
9.73×10^{-3}	cubic meter, m^3	acre-inch	102.8
35.3	cubic meter, m^3	cubic foot, ft^3	2.83×10^{-2}
6.10×10^4	cubic meter, m^3	cubic inch, in^3	1.64×10^{-5}
2.84×10^{-2}	liter, L (10^{-3} m^3)	bushel, bu	35.24
1.057	liter, L (10^{-3} m^3)	quart (liquid), qt	0.946
3.53×10^{-2}	liter, L (10^{-3} m^3)	cubic foot, ft^3	28.3
0.265	liter, L (10^{-3} m^3)	gallon	3.78
33.78	liter, L (10^{-3} m^3)	ounce (fluid), oz	2.96×10^{-2}
2.11	liter, L (10^{-3} m^3)	pint (fluid), pt	0.473

		Mass	
2.20×10^{-3}	gram, g (10^{-3} kg)	pound, lb	454
3.52×10^{-2}	gram, g (10^{-3} kg)	ounce (avdp), oz	28.4
2.205	kilogram, kg	pound, lb	0.454
0.01	kilogram, kg	quintal (metric), q	100
1.10×10^{-3}	kilogram, kg	ton (2000 lb), ton	907
1.102	megagram, Mg (tonne)	ton (U.S.), ton	0.907
1.102	tonne, t	ton (U.S.), ton	0.907
		Yield and Rate	
0.893	kilogram per hectare, kg ha^{-1}	pound per acre, lb $acre^{-1}$	1.12
7.77×10^{-2}	kilogram per cubic meter, kg m^{-3}	pound per bushel, lb bu^{-1}	12.87
1.49×10^{-2}	kilogram per hectare, kg ha^{-1}	bushel per acre, 60 lb	67.19
1.59×10^{-2}	kilogram per hectare, kg ha^{-1}	bushel per acre, 56 lb	62.71
1.86×10^{-2}	kilogram per hectare, kg ha^{-1}	bushel per acre, 48 lb	53.75
0.107	liter per hectare, L ha^{-1}	gallon per acre	9.35
893	tonne per hectare, t ha^{-1}	pound per acre, lb $acre^{-1}$	1.12×10^{-3}
893	megagram per hectare, Mg ha^{-1}	pound per acre, lb $acre^{-1}$	1.12×10^{-3}
0.446	megagram per hectare, Mg ha^{-1}	ton (2000 lb) per acre, ton $acre^{-1}$	2.24
2.24	meter per second, m s^{-1}	mile per hour	0.447
		Specific Surface	
10	square meter per kilogram, m^2 kg^{-1}	square centimeter per gram, cm^2 g^{-1}	0.1
1000	square meter per kilogram, m^2 kg^{-1}	square millimeter per gram, mm^2 g^{-1}	0.001
		Density	
1.00	megagram per cubic meter, Mg m^{-3}	gram per cubic centimeter, g cm^{-3}	1.00
		Pressure	
9.90	megapascal, MPa (10^6 Pa)	atmosphere	0.101
10	megapascal, MPa (10^6 Pa)	bar	0.1
2.09×10^{-2}	pascal, Pa	pound per square foot, lb ft^{-2}	47.9
1.45×10^{-4}	pascal, Pa	pound per square inch, lb in^{-2}	6.90×10^3

(continued on next page)

Conversion Factors for SI and non-SI Units

To convert Column 1 into Column 2, multiply by	Column 1 SI Unit	Column 2 non-SI Units	To convert Column 2 into Column 1, multiply by
		Temperature	
1.00 (K – 273)	kelvin, K	Celsius, °C	1.00 (°C + 273)
(9/5 °C) + 32	Celsius, °C	Fahrenheit, °F	5/9 (°F – 32)
		Energy, Work, Quantity of Heat	
9.52×10^{-4}	joule, J	British thermal unit, Btu	1.05×10^{3}
0.239	joule, J	calorie, cal	4.19
10^{7}	joule, J	erg	10^{-7}
0.735	joule, J	foot-pound	1.36
2.387×10^{-5}	joule per square meter, J m^{-2}	calorie per square centimeter (langley)	4.19×10^{4}
10^{5}	newton, N	dyne	10^{-5}
1.43×10^{-3}	watt per square meter, W m^{-2}	calorie per square centimeter minute (irradiance), cal cm^{-2} min^{-1}	698
		Transpiration and Photosynthesis	
3.60×10^{-2}	milligram per square meter second, mg m^{-2} s^{-1}	gram per square decimeter hour, g dm^{-2} h^{-1}	27.8
5.56×10^{-3}	milligram (H_2O) per square meter second, mg m^{-2} s^{-1}	micromole (H_2O) per square centimeter second, µmol cm^{-2} s^{-1}	180
10^{-4}	milligram per square meter second, mg m^{-2} s^{-1}	milligram per square centimeter second, mg cm^{-2} s^{-1}	10^{4}
35.97	milligram per square meter second, mg m^{-2} s^{-1}	milligram per square decimeter hour, mg dm^{-2} h^{-1}	2.78×10^{-2}
		Plane Angle	
57.3	radian, rad	degrees (angle), °	1.75×10^{-2}

	Electrical Conductivity, Electricity, and Magnetism		
10	siemen per meter, S m^{-1}	millimho per centimeter, mmho cm^{-1}	0.1
10^4	tesla, T	gauss, G	10^{-4}
	Water Measurement		
9.73×10^{-3}	cubic meter, m^3	acre-inch, acre-in	102.8
9.81×10^{-3}	cubic meter per hour, m^3 h^{-1}	cubic foot per second, ft^3 s^{-1}	101.9
4.40	cubic meter per hour, m^3 h$^-$1	U.S. gallon per minute, gal min^{-1}	0.227
8.11	hectare meter, ha m	acre-foot, acre-ft	0.123
97.28	hectare meter, ha m	acre-inch, acre-in	1.03×10^{-2}
8.1×10^{-2}	hectare centimeter, ha cm	acre-foot, acre-ft	12.33
	Concentrations		
1	centimole per kilogram, cmol kg^{-1}	milliequivalent per 100 grams, meq 100 g^{-1}	1
0.1	gram per kilogram, g kg^{-1}	percent, %	10
1	milligram per kilogram, mg kg^{-1}	parts per million, ppm	1
	Radioactivity		
2.7×10^{-11}	becquerel, Bq	curie, Ci	3.7×10^{10}
2.7×10^{-2}	becquerel per kilogram, Bq kg^{-1}	picocurie per gram, pCi g^{-1}	37
100	gray, Gy (absorbed dose)	rad, rd	0.01
100	sievert, Sv (equivalent dose)	rem (roentgen equivalent man)	0.01
	Plant Nutrient Conversion		
	Elemental	***Oxide***	
2.29	P	P_2O_5	0.437
1.20	K	K_2O	0.830
1.39	Ca	CaO	0.715
1.66	Mg	MgO	0.602

1 Multidimensional Thinking: A Prerequisite to Agroecology

DIANE RICKERL

South Dakota State University
Brookings, South Dakota

CHARLES FRANCIS

University of Nebraska
Lincoln, Nebraska

Agroecology employs the principles of ecology to study, understand, and design agricultural systems. One of the most basic principles of ecology is that changing one component changes everything else. Research and teaching in agricultural science is evolving from using only the conventional cause and effect or linear mentality to a holistic approach focusing on sustainability. A holistic orientation requires a more comprehensive way of looking at things, a more multidimensional view. A thought system such as multidimensional thinking is not linear, or even circular, but spherical, including a time dimension. The evolution to this approach is the result of a natural progression in understanding complexity, and even a logical developmental process. Such thinking provides a framework or context into which component technologies can be organized. In this chapter we take up the important task of defining multidimensional thinking. We provide examples from different areas of study, or windows on the agricultural system, that we call disciplines. In conclusion, we discuss multidimensional thinking as it relates to agroecology.

WHAT IS MULTIDIMENSIONAL THINKING?

Multidimensional thinking is difficult to achieve in our discipline-specific departments. In some ways we are like the six blind fellows in the poem by John Godfrey Saxe, "The Blind Men and the Elephant" (Untermeyer, 1963). Each suffers from the same disability (i.e., each is in a specific discipline), and each approaches a different part of the elephant (i.e., specialization). The poem points out that reductionist thinking, even by wise people, does not often give us the complete picture. Each of the wise people is partly in the right, but all are in the wrong in terms of the larger structure and function. These specialists demonstrate that the description of the parts cannot possibly give the whole picture and that linear thinking is sometimes inadequate. The large creature we are trying to understand is agriculture, and the innovative, multidimensional approach described in this book is agroecology.

Multidimensional thinking may be considered a prerequisite to the development of agroecology. A narrow focus helps us become more specialized and productive in our disciplines, yet we may have narrowed our field of vision until we become like the blind men in the poem with respect to wider issues in agriculture. Since we often relate dimension to the visual arts, construction projects, or geometry, it is useful to begin with several visual, concrete examples. First, let us consider some regional artists from the Midwest.

Czestochowski (1981) described Grant Wood as an artist who "gave expression to a living experience" and whose "art narrates the frenzied transformation in American values from those of a rural community to those of an urban-industrial society...reaffirming the uniqueness of American culture as a social and environmental phenomenon." Wood mastered the illusion of dimension. It is easy to see the differences in dimension among Grant Wood's paintings. *Plowing on Sunday*, with only the farmer in the foreground, is unidimensional (Fig. 1–1); *American*

Fig. 1–1. Grant Wood (American, 1891–1942), *Plowing on Sunday*, 1934. Copyrighted by the Estate of Grant Wood. Licensed by VAGA, New York. Museum of Art, Rhode Island School of Design, Gift of Mrs. Murray S. Danforth.

Gothic is a two-dimensional presentation of the farmer and his wife in the foreground and a farm house with a Gothic window in the background (Fig. 1–2); and *Stone City* (Fig. 1–3) is a three-dimensional view that carries the eye for miles across the Iowa countryside. These differences in dimension are perceptions created by the artist.

Through relief or sculpture, visual artists can further achieve multidimensionality by creating works that exist in space. Christian Petersen's sculpture *The Cornhusker* (Fig. 1–4) depicts a young man from Nevada, IA who has won the county, district, and state husking championships. Bliss (1986) describes the sculpture as "Petersen's tribute to a uniquely Midwestern type of athlete in the pre-

Fig. 1–2. Grant Wood (American, 1891–1942), *American Gothic*, 1930. Oil on beaverboard, 74.3 by 62.4 cm. Friends of American Art Collection. All rights reserved by The Art Institute of Chicago and VAGA, New York, NY, 1930.9341, reproduction copyrighted by The Art Institute of Chicago.

Fig. 1–3.Grant Wood (American, 1891–1942), *Stone City, Iowa*, 1930. (JAM.1930.35), Joslyn Art Museum, Omaha, NE.

mechanized farm era." South Dakota has one of the nation's most famous relief sculptures, Mount Rushmore. The viewer is consumed by the immense lifelike faces of four presidents. An even more immense mountain sculpture of Crazy Horse is moving toward completion a few miles to the south. The use of real physical dimension, and in these examples, scale, adds to their impact. Their historical relevance adds a fourth dimension, time.

In the 1960s, sculptors began to create environments as their three-dimensional work. The impact of works such as *The Farm Worker* (Fig. 1–5) by George Segal, which includes life-size sculptures in a created space, comes from the total environment created by the artist. Segal (Seitz, 1972) commented about one of his works, "I was responding to all the polarities, and had to find a way to connect real space, direct sensation, with a metaphysical dream state. Specific and general had to become entangled; figure and abstraction had to fuse."

In the study of geometry, dimension is defined and drawn. Students first learn the definition of a point, then a line, a plane, and a solid. The concept of volume is not taught before the concept of area, as that would not be a logical progression. For most of us, multidimensional thinking has progressed through our personal development and education, just as we learned the details of geometry. In contrast, most courses packaged within disciplines require us to pull back into thinking about components of systems, and thus address challenges in fewer dimensions. It is difficult to assemble the pieces of a puzzle if each has been modified or refined without doing this within the context of the larger picture.

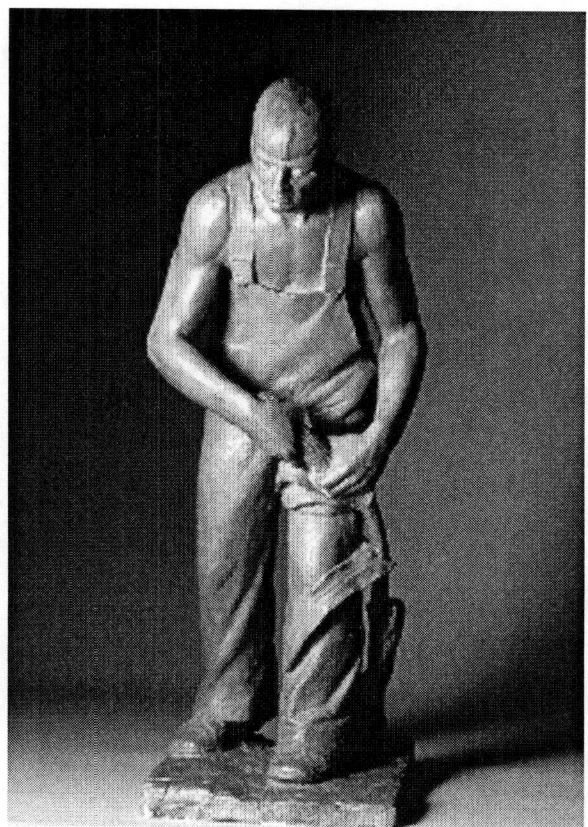

Fig. 1–4. Christian Petersen (American, 1885–1961), *Cornhusker*, 1941. Painted plaster, 43 by 16 by 24 inches. Christian Petersen Collection, Brunnier Art Museum. University Museums, Iowa State University. Gift in memory of Joseph M. Coppola, Sr., by the Coppola Family. UM99.329.

In Piaget's studies (1973) of intellectual development of children he states, "I will call it psychological, the development of the intelligence itself—what the child learns by himself [*sic*], what none can teach him and he must discover alone; and it is essentially this development which takes time." Piaget describes sequential stages of development that proceed from concepts of substance as an empty form, to conservation of weight, and finally to conservation of volume. This developmental progression is analogous to the concepts of point, line, plane, and volume in the principles of geometry. However, agricultural systems are far more complex and involve the interaction of many biological and physical factors, as well as human ambitions and survival within the natural and the built environments. Kirshenmann (2004) tells us in Chapter 11 of this book, Ecological Morality, that philosophers, sages, and shamans have understood that humans are part of a complex web of life. Our only chance to understand how systems function, and how our component data fit in, is to embrace this complexity, and for some of us to look carefully at whole systems.

Conceptual development is often expressed visually in the progression of a child's drawings, going from dots and slashes, to enclosed shapes, to recognizable forms (Fig. 1–6). All of us have seen the lack of perspective (dimension) in children's drawings where trees and houses sit on the horizon line. Even projects built from modeling clay (a three-dimensional media) often take on a two-dimensional quality in the hands of a young child. Piaget summarizes:

> This order of succession shows that, if a new instrument of logic is to be constructed, there must always be previous logical instruments, that is the construction of a new

Fig. 1–5. George Segal, *The Farm Worker*, 1962–1963. Copyrighted by The George and Helen Segal Foundation. Licensed by VAGA, New York.

> notion will always suppose substrata, previous substructures, and this by indefinite regressions.
>
> (Piaget, 1973)

The medium, modeling clay, cannot be used in three-dimensional expressions until the concept of volume has been developed. To suggest that research on component technology is more "childlike" than that of system's integrators would be highly erroneous—in fact it may be the system's integrators who need to revisit this concept of how the appreciation of spatial dimension develops. These integrators have the responsibility to build effectively on available component research and technologies, and to explore the complexities of interactions and emergent properties of systems at several levels of hierarchy. In fact, those unencumbered by heavy, component-oriented detail may be better able to grasp the larger issues and integrate components into creative new systems.

Our understanding of geography is an excellent example of the progression of multidimensional thinking. By the time of Christopher Columbus there had been a great deal of change in perceptions about the earth. The local "point" focus of pre-

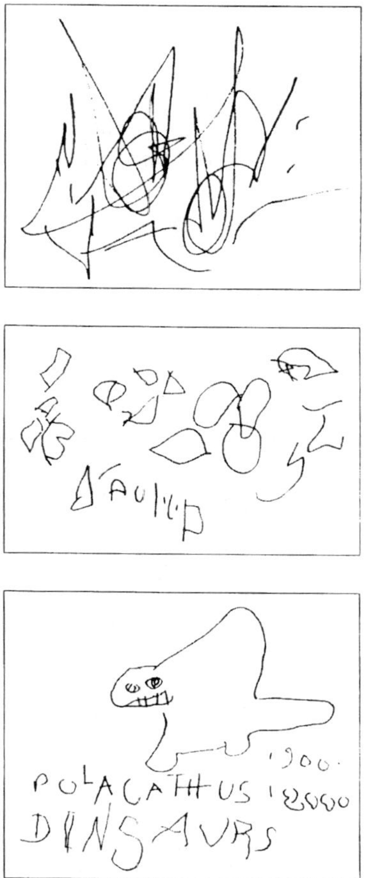

Fig. 1–6. Children's drawings showing the progression in conceptual development.

historic humans in their tribal territories had progressed to a broader two-dimensional theory of a flat earth. With some evidence and multidimensional hopes, Columbus began his voyage across the ocean. Even though he didn't know where he was going, or where he had been when he returned, Columbus's dimensional perception had been enhanced. Ideas began to change. Similarly, in 1969 when Neil Armstrong walked on the moon as the world watched, multidimensional thinking took a giant step through both space and time.

Leonardo da Vinci was an Italian artist and scientist, one of the first "renaissance men" and an excellent example of a person of multidimensional thought. Best known for his classical paintings, Leonardo was also an architect, engineer, anatomist, and sculptor. Today's university would have difficulty finding a department or even a college where his immense talents would fit into the confines of one discipline.

In biological science, an example of multidimensional thinking can be illustrated through advances in the development of the microscope. Robert Hooke writes in *Micrographia* (Espinasse, 1956) "...these pores, or cells, were the first microscopical pores I ever saw, and perhaps that were ever seen...." His discovery was named the "cell" because only the lines of the cell wall were visible using his microscope (Fig. 1–7). Hooke was a multidimensional thinker. He was accomplished in science, art, and engineering. Hooke's friend Wren (Espinasse, 1956) wrote, "of him I must affirm, that, since the time of Archimedes, there scarce ever met in one man, in so great a perfection, such a Mechanical Hand and so Philosophical a Mind."

With further development, the microscope revealed cell contents as well as cell walls (Fig. 1–8). The scanning electron microscope added the third and fourth dimensions to microscopy (Fig. 1–9). Today we can watch cells function and follow their life cycles through time using technology that has progressed from unidimensional to three- and then four-dimensional. As in the example of Piaget's theory of intellectual development, each phase in the development of the microscope was built on the previous technology and understanding and would have been impossible without it.

Computer technology is perhaps today's most rapidly changing example. Computers initially processed numerical data. Before long they could use instructions to plot or chart data, then surfaces and moving three-dimensional images. Virtual realities generated by computers today can take us through experiences that are multidimensional. In this case a sense of reality is gained from the combination of visual, audio, and tactile experience. Perhaps we should be grateful that development of computer imagery has been spurred by the military and by action video games and that these advances may soon be applied for more beneficial uses in society. Technology has allowed the enhancement of sensory perception. It is the illusion of an environment, the same concept used by Segal.

These examples of the progression of multidimensional thinking indicate changes in perception through time. Our perception of time itself has also evolved. In the ancient Mayan culture it was believed that humans were placed on earth to bear the burden of time, to be the timekeepers. Einstein in his theory of relativity regarded the time variable as a Euclidean four-dimensional continuum. In the words of Niels Bohr (Clark, 1971), "Mankind will always be indebted to Einstein for the removal of the obstacles to our outlook which were involved in the primitive notions of absolute space and time." The Mayans, Columbus, Neil Armstrong, Hooke, and Einstein are just a few examples of multidimensional thinkers. Why are they the exceptions?

There are many obstacles to multidimensional thinking. We have discussed limitations in technology, society, university department organization, and self. Microscopes and computers have changed our perceptions and given us new insights. Societies and their religions have been obstacles to creative and multidimensional thinking. Popular belief in the 1400s was that the world was flat. Columbus' theories were rejected in Portugal and initially in Spain. Only his persistence led to his success. When Einstein proposed that the energy contained in a light beam is transferred in quanta, he contradicted a century-old theory. Almost no one accepted the

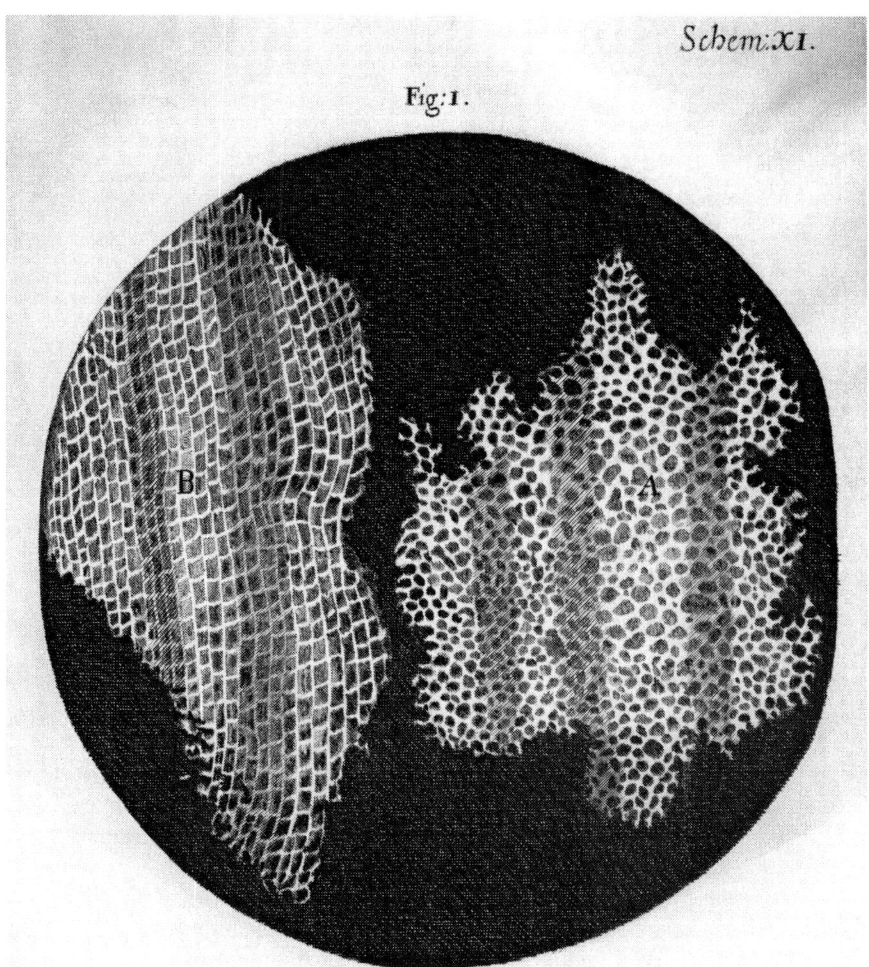

Fig. 1–7. Robert Hooke's representation of cork cells, *Micrographia: Or Some Physiological Descriptions of Minute Bodies Made by Magnifying Glasses*, London, 1665. Courtesy of Smithsonian Institution Libraries, Washington, DC.

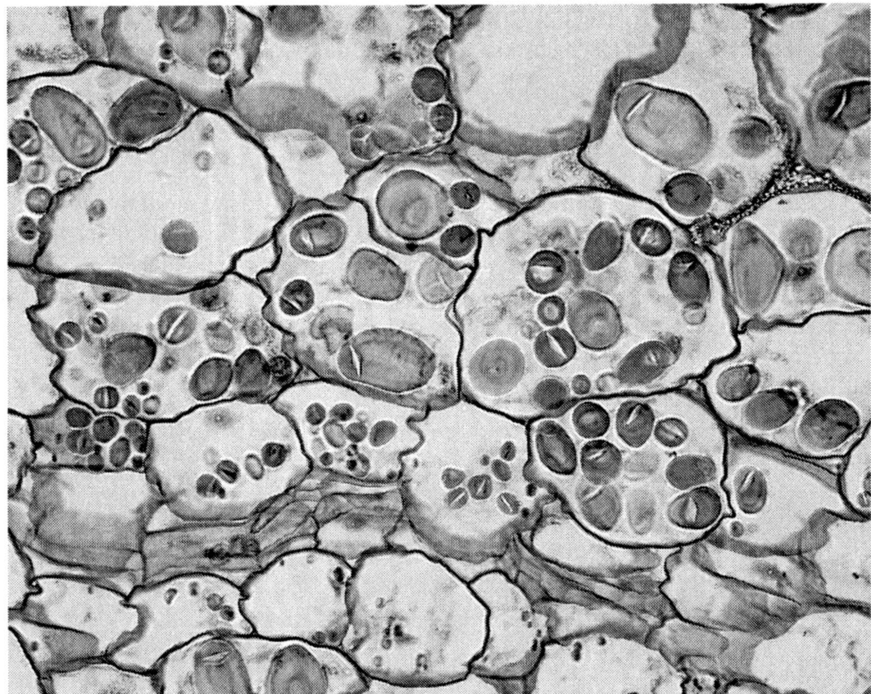

Fig. 1–8. Transmission electron micrograph of potato cells with starch granules. Courtesy of Michael W. Davidson.

new proposal. Even when Robert Andrews Miliken experimentally confirmed the theory a decade later, he was surprised and uncomfortable with the outcome. Darwin was extremely cautious in explaining his theories of evolution to the public, often sitting on his results for five years or more. He would have found difficulty making tenure in universities today. The views of society, including the university community, can be a strong limitation, but society can progress, just as the individual progresses through developmental stages.

Thomas Kuhn (1996) took the argument that society can impede progress further when he suggested that evidence contrary to our predominant paradigms can be virtually invisible to the observer. This is why change occurs on the fringes, and rarely in the mainstream (Barker, 1985). Rachel Carson's (1962) *Silent Spring* attacked the widespread use of pesticides and precipitated a radical change in agriculture, but this was only possible because of her biology background and her ability to see the larger picture.

Agroecology requires multidimensional thinking, the ability to see the whole elephant as well as the context in which it functions. The examples given illustrate the importance of an individual's psychological development in his/her ability to think in multiple dimensions. The importance of society and the prevailing views are also instrumental in shaping science and our understanding of systems. Technology can offer tools to speed the development of dimensional thinking, but in it-

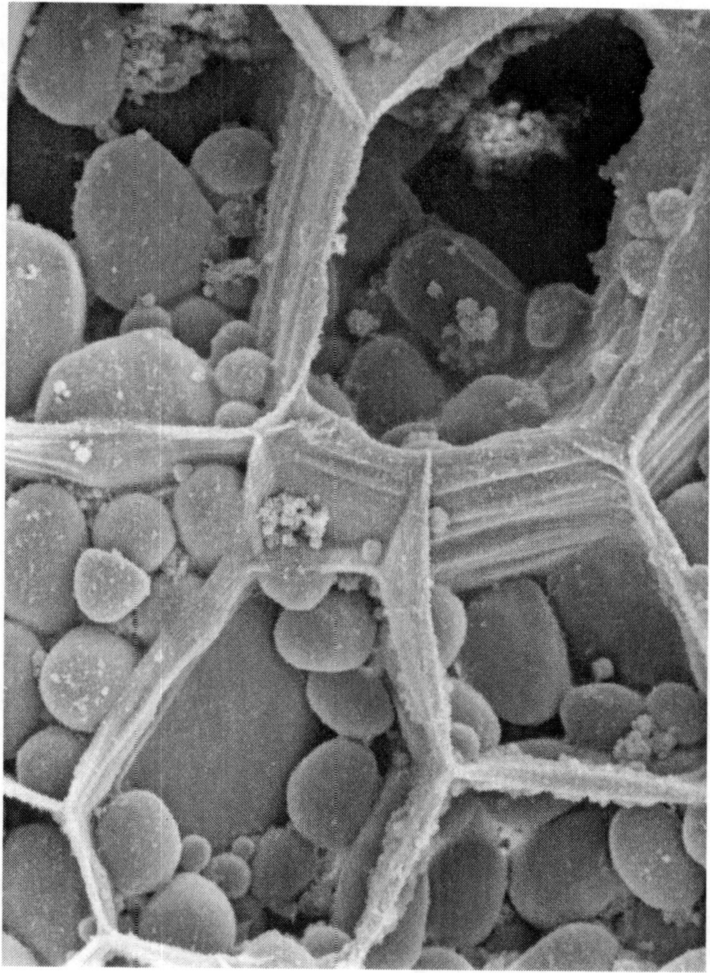

Fig. 1–9. Scanning electron micrograph of potato cell with starch granules. Courtesy of Institute of Food Research, United Kingdom.

self it is not necessarily multidimensional. It takes a unique individual or society to defy the current theories of culture and science to pursue different views.

PURSUING A MULTIDIMENSIONAL VIEW OF AGROECOLOGY

Agroecology has been defined variously as the marriage of agriculture and ecology, the understanding of crop ecology in the context of cultivated systems, and the description of agriculture in ecological terms. Gliessman (1998) traced the history of uses of this term back to the early part of the 20th century, when field ecology and practical agronomy shared common interests in cultivated and natural systems. Crop ecology was an important area of study until ecologists moved toward greater focus on natural systems, and agronomists became more specialized in disciplines related to specific components of the system. The Gliessman chronology has been expanded to include more references from Latin America and Europe in a recent article on defining agroecology as the ecology of food systems (Francis et al., 2003).

What are the dimensions contained in an agroecological view and what approaches can be taken to reach the broad and needed level of understanding? Okigbo (1989) listed five elements that should always be considered to achieve a sustainable food system in a given area, to which we add a sixth dimension related to the environment:

- physiochemical factors—soils, climate, moisture, radiation, day length
- biological elements—crops animals, pests
- changing and appropriate technologies available to the farmer
- sociocultural background—education, policy, experience
- economic viability—market, costs, management
- ecological soundness—preservation of biodiversity and ecosystem functions

Okigbo emphasized that these dimensions should each be considered for sustainable development at the global, regional, national, and local levels. Further discussion of the importance of hierarchy and scale is found in Chapter 2 (Gliessman, 2004).

Harwood's (1990) history of sustainable agriculture suggests two reference points for the evolution of this concept. The first is "agriculture based on principles of ecological interaction," which emerged in the early 1980s, and the second is "stable agriculture in the global sense, involving all facets of agriculture and its interaction with society," which emerged later in the decade. Harwood's own definition, given certain limitations, is "an agriculture that can evolve indefinitely toward greater human utility, greater efficiency of resource use, and a balance with the environment that is favorable both to humans and to most other species." The definition incorporates values, but allows for agricultural development. Harwood reviews a thought development process (multidimensional thinking) of major significance to the concept of sustainability. From this process, stemming from the early 1900s, there emerge

- the interrelatedness of all parts of a farming system, including the farmer and his family
- the importance of the many biological balances in the system
- the need to maximize desired biological relationships in the system and to minimize use of material and practices that disrupt those relationships

In the late 1900s, as communication and transportation technology enabled global exchange, we began to realize the role of human diversity and community in food systems. Butler and DePhelps (1995) discussed the development of rural and urban communities in America. Their sequence fits the progression from point to sphere to time in multidimensional thinking. The point is characterized by many small rural communities located in a specific geographic place. Technology allowed interactions between communities with shared interests (linear) as rural communities began to decline. In the 1990s bioregionalism (circular) emphasized the interconnectedness within a geographic region and the idea that cultural identity is connected to the natural environment.

> In a sense, bioregionalism is a shift of emphasis from national or global to local communities within a specific geographic region. This shift does not ignore the relationships between local and global spheres. Rather, it focuses energy within local communities while strengthening relationships among communities, within regions and beyond.
>
> (Butler & DePhelps, 1995)

In this view, bioregionalism can become fully dimensional, looking at relationships among interacting spheres. The bioregion becomes a common ground for effective community-based problem solving (multidimensional thinking). Further, the relationships among these communities develop and evolve with time, changing in response to demographics, resources, and political circumstances among other factors.

To encourage community member participation a well-developed sense of place is needed. "To know the spirit of a place is to realize that you are part of a part and that the whole is made of parts, each of which is whole. You start with the part you are whole in." Butler and DePhelps (1995) offered approaches for developing what is essentially an agroecological view for community-based problem solving. They suggested techniques and provided examples of strategies that encourage open dialog and create trust among community members.

In a recent commentary article we proposed a broader definition of agroecology as the "ecology of food systems" (Francis et al., 2003). We suggested that agroecology is an appealing term that helps scientists, farmers, and others focus on both the structure and function of agroecosystems, including the long-term positive and negative impacts of systems on humans and other species. Although the focus in agricultural science and current agroecology courses has been primarily on the production components and environmental impacts, Francis and colleagues argued that attention to processing, marketing, and consumption dimensions will help our understanding of the total energy and materials flows in the system and lead to design of more efficient resource use and human utility in the systems. One goal of their article is to stimulate teachers and researchers in agroecology to embrace this broader concept of studying the total food system, and thus to add multiple dimensions to our understanding of how systems function and how they can be improved.

POTENTIAL FOR AGROECOLOGY IN UNIVERSITIES

The results of a survey of land grant universities conducted by Francis et al. (1995) suggest that, "There is great potential for cooperative ventures among universities to address both the sustainability of agriculture and the concerns of society." Although most of the current business as usual takes place within the confines of discipline-specific departments, there are multidiscipline centers and areas of concentration in teaching that demonstrate how new approaches can be implemented within the current system. A list of potential future directions focuses on forming and strengthening coalitions and improving communication networks that bridge the traditional boundaries and create new communities of scholars who can deal with complex issues.

Student demand for education in agricultural systems in the USA and in ecological agriculture in the Nordic Region has stimulated the search for new approaches to organizing learning. Lieblein et al. (2000) presented a critique of the current university structure that places all activities in boxes that correspond to courses in specific topics, too often taught in isolation from other courses and from the operating world of agriculture. They pointed out that today's society is beginning to demand that ecological, ethical, and social dimensions must be included in our design of food systems and in shaping the future of agriculture. If the agricultural universities are unable to respond to the need for students to understand complexity and the multiple dimensions of tomorrow's food systems, it is likely that other players will enter the educational arena (Kunkle et al., 1996).

Lieblein et al. (2000) suggested that universities' curricula are based on the assumption that there is a large gap between ignorance and knowledge, while in reality there is a much larger gap today between knowing and doing. For this reason, they recommended active learning based on student projects and greater emphasis on activities outside the classroom. Their concept of a "future learning university" is shown in Fig. 1–10. An expanded definition of faculty, which includes people from business, government, and farming, can enhance the classroom education, as it recognizes that students learn in many ways and in different settings. (Francis et al., 2003). These concepts are being introduced in the Nordic graduate programs in agroecology and ecological agriculture, the term used in that region to describe organic farming.

MULTIPLE DIMENSIONS OF AGROECOLOGY

This book explores the many dimensions of agricultural systems that are needed to fully understand how they are structured and how they work. It is convenient to approach the system at different spatial levels in order to understand the dynamics of biological and geochemical reactions that occur there. It is essential to focus on the foundation of soil and climate to understand how these set limits on potential productivity of systems. Equally important are the human participants in the system and how they design and manage resources to extract food and other products from the agroecosystem. It is the multidimensional view of these different levels that make this book unique.

The types of structures and the processes that occur at different levels in the spatial hierarchy are described in Chapter 2. The functions of an agroecosystem cannot be well understood unless this is viewed from different points of view, and this requires a multidisciplinary approach and analysis of system functions. Soil dynamics and plant nutrition are at the foundation of system potential, and our understanding of these processes is growing as we study the biological complexities of soil organisms and nutrient interactions unconfounded by massive applications of fertilizers and pesticides (Chapter 3). Insects, weeds, and pathogens interact with

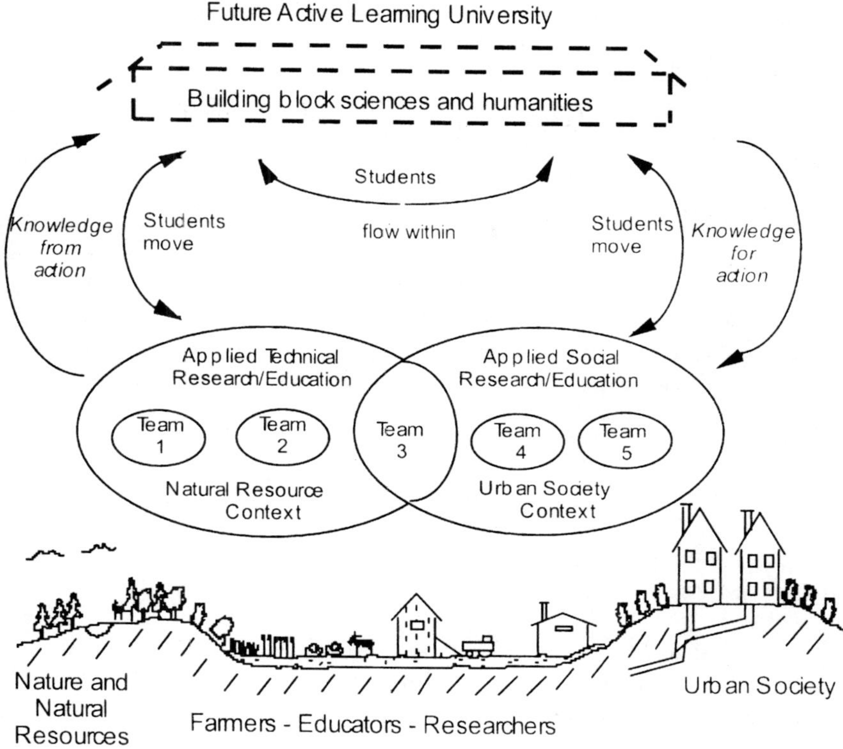

Fig. 1–10. Schematic description of future university for active learning and research, with student and faculty learning activities often in the farming environment, and close relationships with both natural resource and urban society contexts (from Lieblein et al., 2000; reprinted with permission of *Journal of Agricultural Education & Extension*, Wageningen, the Netherlands).

crops and animals at the field and farm levels to influence success of the agricultural enterprises, and the complexity of these processes is discussed in Chapter 4. The aggregation of functions of individual crop plants and domestic animals leads to study of the whole farm, and how its dynamic nature can be directed through alternative approaches to management (Chapter 5).

Economics can be used to evaluate agricultural systems in multiple ways. The neoclassical approach to economic analysis is but one method of evaluating success, and a multifaceted approach to measurement is presented in Chapter 6. The importance of social capital and the dynamics of rural communities are discussed in their many dimensions in Chapter 7. How the development of systems contributes or detracts from the conservation of resources is presented in Chapter 8. Global information systems are extremely valuable in assessing the spatial variability of many dimensions of agricultural systems and with time can provide a dynamic picture of how these change (Chapter 9).

An emerging interest in agriculture is the multifunctionality of rural landscapes and how society recognizes and rewards the ecological services these landscapes provide. Methods of measuring ecological functions are presented in Chapter 10. Including humans and their activities as key drivers in the design and function of agricultural systems brings in ethical dimensions of agriculture, including how we treat the landscape, its resources, and each other. These are discussed in Chapter 11. The future of agroecology as it relates to better understanding and design of food systems is explored in Chapter 12, including a summary of the conclusions from the preceding chapters.

The evolution of agriculture in America brings us to a time that is right for using the concepts and tools that are emerging in our expanding field of agroecology. The complex problems of local and regional food shortages and environmental degradation which face our global society will require approaches that integrate the environment, society, and ecology. We are being asked to develop a new ethic for agriculture, an ecological morality that will reflect our understanding of nature. Community-based problem solving and change within the land grant universities are approaches that can be used. Regardless of the approach, our way of thinking has to be multidimensional.

STUDY QUESTIONS

1. How can multidimensional thinking be applied to solving water pollution problems caused by excessive pesticide application in agriculture?

2. This chapter describes a progression from simple to complex dimensional expression in paintings. What similar progression or parallel can be found in music, in theater, in architecture, or in other artistic expressions?

3. Identify and describe some examples in your own discipline where there have been quantum leaps in multidimensional thinking as a result of discoveries of new methods or tools for analysis?

4. Discuss the advantages and disadvantages of defining agroecology broadly as the ecology of food systems.

REFERENCES

Barker, J.A. 1985. Discovering the future: The business of paradigms. ILI Press, Lake Elmo, MN.

Bliss, P.L. 1986. Christian Petersen remembered. Iowa State Univ. Press, Ames.

Butler, L.M., and C. DePhelps. 1995. Human diversity, community, and viable food and agricultural systems. p. 161–193. *In* R.K. Olson et al. (ed.) Exploring the role of diversity in sustainable agriculture. ASA, Madison, WI.

Carson, R. 1962. Silent spring. Houghton Mifflin, Boston.

Clark, R.W. 1971. Einstein, the life and times. The World Publishing Company, New York.

Espinasse, M. 1956. Robert Hooke. Univ. of California Press, Berkeley.

Czestochowski, J.S. 1981. John Steurat Curry and Grant Wood: A portrait of rural America. Univ. of Missouri Press, Columbia.

Francis, C., C. Edwards, J. Gerber, R. Harwood, D. Keeney, W. Liebhardt, and M. Miebman. 1995. Impact of sustainable agriculture programs on U.S. landgrant universities. J. Sustain. Agric. 5(4):19–33.

Francis, C., G. Lieblein, L. Salomonsson, J. Helenius, T.A. Breland, D. Rickerl, N. Creamer, R. Salvador, M. Wiedenhoeft, S. Simmons, P. Allen, S. Gliessman, M. Altieri, J. Porter, R. Harwood,

C. Flora, and R. Poincelot. 2003. Agroecology: The ecology of food systems. J. Sustain. Agric. 22:(in press).

Gliessman, S.R. 1998. Agroecology: Ecological process in sustainable agriculture. Sleeping Bear Press, Chelsea, MI.

Gliessman, S.R. 2004. Agroecology and agroecosystems. p. 19–30. *In* Agroecosystems analysis. Agron. Monogr. 43. ASA, CSSA, SSSA, Madison, WI.

Harwood, R. 1990. A history of sustainable agriculture. p. 3–19. *In* C. Edwards et al. (ed.) Sustainable agricultural systems. Soil & Water Conservation Society, Ankeny, IA.

Kirschenmann, F. 2004. Ecological morality: A new ethic for agriculture. p. 167–176. *In* Agroecosystems analysis. Agron. Monogr. 43. ASA, CSSA, SSSA, Madison, WI.

Kuhn, R. 1996. The structure of scientific revolutions. 3rd ed. Univ. Chicago Press, Chicago.

Kunkle, H.O., I.A. Maws, and C.L. Skaggs (ed.) 1996. Revolutionizing higher education in agriculture. Iowa State Univ. Press, Ames.

Lieblein, G., C. Francis, and J. King. 2000.Conceptual framework for structuring future agricultural colleges and universities in industrial countries. J. Agric. Educ. Exten. 6(4):213–222.

Okigbo, B.N. 1991. Development of sustainable agricultural production systems in Africa: Roles of international agricultural research centers and national agricultural research systems. IITA, Ibadan, Nigeria.

Piaget, J. 1973. The child and reality. Grossman Publishers, New York.

Seitz, W. 1972. Segal. Harry N. Abrams Publ., New York.

Untermeyer, L. 1963. The golden treasury of poetry. Golden Press, New York.

2 Agroecology and Agroecosystems

STEPHEN R. GLIESSMAN

Department of Environmental Studies
University of California
Santa Cruz, California

Agriculture is more than an economic activity designed to produce a crop or to make as large a profit as possible on the farm. A farmer can no longer pay attention to the objectives and goals for his or her farm only and expect to adequately deal with the concerns of long-term sustainability. Discussions about sustainable agriculture must go far beyond what happens within the fences of any individual farm. Farming is now viewed as a much larger system with many interacting parts, including environmental, economic, and social components (Gliessman, 2001; Flora, 2001). It is the complex interaction and balance among all of these parts that has brought us together to discuss sustainability, to determine how to move toward this broader goal, and to learn how an agroecological perspective focused on sustainable agroecosystems is a way to achieve these long-term objectives.

Much of modern agriculture has lost the balance needed for long-term sustainability (Kimbrell, 2002). With their excessive dependence on fossil fuels and external inputs, most industrialized agroecosystems are overusing and degrading the soil, water, genetic, and cultural resources upon which agriculture has always relied. Problems in sustaining agriculture's natural resource foundation can only be masked for so long by modern practices and high input technologies. In a sense, as we borrow ever-increasing amounts of water and fossil fuel resources from future generations, the negative impacts on farms and farming communities will continue to become more evident. The conversion to sustainable agroecosystems must become our goal (Gliessman, 2001).

In an attempt to clarify my own thinking about agroecosystems, I often think of agriculture as a stream, and farms are different points along that stream. When we think of an individual farm as a "pool" in a calm eddy at some bend in the stream's flow, we can imagine how many things "flow" into a farm, and we also expect that many things flow out of it as well. As a farmer, I work hard to keep my pool in the stream (my farm) clean and productive. I try to be as careful as possible in terms of how I care for the soil, which crops I plant, how I control pests and diseases, and how I market my harvest. Back in the days when there were fewer farms, fewer people to feed, and smaller demands on farmers and farmland, I could keep my farm in pretty good shape. I could keep my pool in the stream pretty clean and did not have to worry very much about what was going on "downstream" from my farm.

But such a strategy has become much more difficult today. I find that I have less and less control over what comes into my pool. I face a variety of "upstream impacts" that in combination can threaten the sustainability of my farm. These include the inputs into my farm that either I purchase or which arrive from the surrounding area. They include labor availability and cost, market access for what I produce, legislated policies that determine how much water I use, pesticides I apply, or how I care for my animals—not to mention the vagaries of the weather! My pool can become quickly muddied.

I must also increasingly consider how the way I take care of my pool can have "downstream effects" in the stream below. Soil erosion and groundwater depletion can negatively affect other farms than my own. Inappropriate or inefficient use of pesticides and fertilizers can contaminate the water and air, as well as leave potentially harmful residues on the food that my family and others will consume. How well I do on my farm is reflected in the viability of rural farm economies, our local community, and society broadly. Key indicators are the losses of farmland to other activities and the loss of family farms in general. Both upstream and downstream factors are linked in complex ways, often beyond my control, and they impinge upon the sustainability of my farm.

THE AGROECOLOGY PERSPECTIVE

The Agroecosystem

Any definition of sustainable agriculture must include how we examine the production system as an agroecosystem. We need to look at the entire system, the entire stream in the above analogy. This definition must move beyond the narrow view of agriculture that focuses primarily on the development of practices or technologies designed to increase yields and improve profit margins. These practices and technologies must be evaluated on their contributions to the overall sustainability of the farm system. The new technologies have little hope of contributing to sustainability unless the longer-term, more complex impacts of the entire agricultural system are included in the evaluation. The agricultural system is an important component of the larger food system (Francis et al., 2003).

A primary foundation of agroecology is the concept of the ecosystem, defined as a functional system of complementary relations between living organisms and their environment, delimited by arbitrarily chosen boundaries, which in space and time appears to maintain a steady yet dynamic equilibrium (Odum, 1996; Gliessman, 1998). Such an equilibrium can be considered to be sustainable in a definitive sense. A well-developed, mature natural ecosystem is relatively stable, self sustaining, recovers from disturbance, adapts to change, and is able to maintain productivity through using energy inputs of solar radiation alone. When we expand the ecosystem concept to agriculture and consider farm systems as agroecosystems, we have a basis for looking beyond a primary focus on traditional and easily measured system outputs (yield or economic return). We can instead look at the complex set of biological, physical, chemical, ecological, and cultural interactions determining the processes that permit us to achieve and sustain yields.

Agroecosystems are often more difficult to study than natural ecosystems because they are complicated by human management, which alters normal ecosystem structures and functions. There is no disputing the fact that for any agroecosystem to be fully sustainable, a broad series of interacting ecological, economic, and social factors and processes must be taken into account. Still, ecological sustainability is the building block upon which other elements of sustainability depend.

An agroecosystem is created when human manipulation and alteration of an ecosystem take place for the purpose of establishing agricultural production. This introduces several changes in the structure and function of the natural ecosystem (Fig. 2–1) and resulting changes in a number of key system-level qualities. These qualities are often referred to as the emergent qualities or properties of systems, qualities that manifest themselves once all of the component parts of the system are organized. These same qualities can also serve as indicators of agroecosystem sustainability (Gliessman, 2001). Four key emergent qualities of ecosystems and how they are altered as they are converted to agroecosystems are discussed in the following sections.

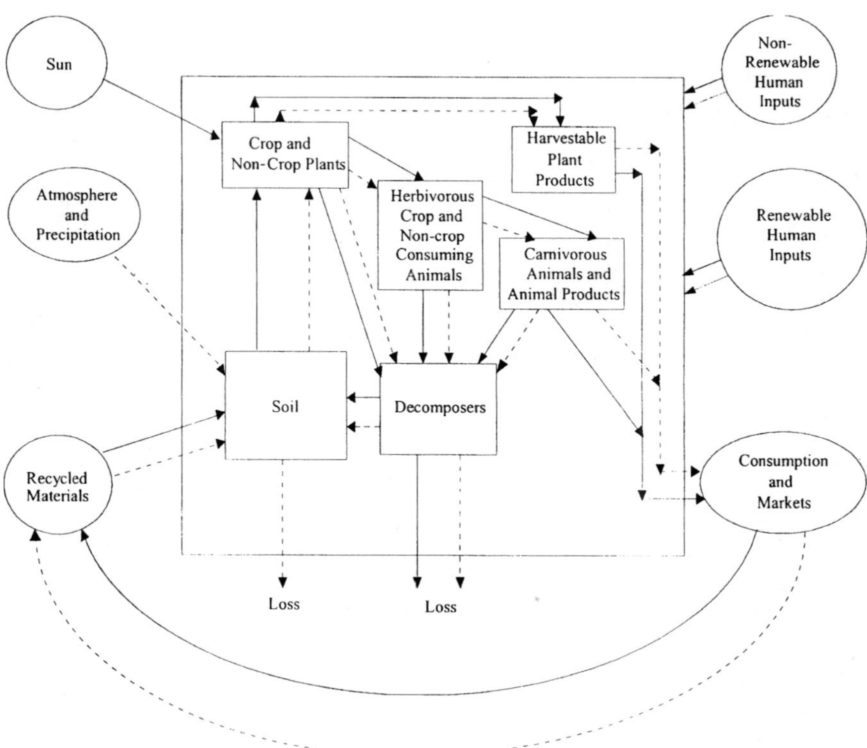

Fig. 2–1. Functional and structural components of an ecosystem converted to a sustainable agroecosystem. Solid lines are energy flow, and dotted lines are nutrient cycles. This model assumes that nutrients and leftover energy are returned to the agroecosystem as reusable materials, and that the use of nonrenewable human inputs is minimized.

Energy Flow

Energy flows through a natural ecosystem as a result of complex sets of trophic interactions, with certain amounts being dissipated at different stages along the food chain, and with the greatest amount of energy within the system ultimately moving along the detritus pathway (Odum, 1971). Annual production of the system can be calculated in terms of net primary productivity or biomass, each component with its corresponding energy content. Energy flow in agroecosystems is altered greatly by human interference (Rappaport, 1971; Pimentel and Pimentel, 1997). Although solar radiation is obviously the major source of energy, many inputs are derived from human-manufactured sources and are most often not self sustaining. Agroecosystems too often become through-flow systems, with a high level of fossil fuel input and considerable energy directed out of the system at the time of each harvest. Biomass is not allowed to otherwise accumulate within the system or contribute to driving important internal ecosystem processes (e.g., organic detritus returned to the soil serving as an energy source for microorganisms that are essential for efficient nutrient cycling). For sustainability to be attained, renewable sources of energy must be maximized, and energy must be supplied to fuel the essential internal trophic interactions needed to maintain other ecosystem functions.

Nutrient Cycling

Small amounts of nutrients continually enter an ecosystem through several hydrogeochemical processes. Through complex sets of interconnected cycles, these nutrients then circulate within the ecosystem, where they are most often bound in organic matter (Borman and Likens, 1967). Biological components of each system become very important in determining how efficiently nutrients move, ensuring that minimal amounts are lost from the system. In a mature ecosystem, these small losses are replaced by local inputs, maintaining a nutrient balance. Biomass productivity in natural ecosystems is linked very closely to the annual rates at which nutrients are able to be recycled. In an agroecosystem, recycling of nutrients can be minimal, and considerable quantities are lost from the system with the harvest or as a result of leaching or erosion due to a great reduction in permanent biomass levels held within the system (Tivy, 1990). The frequent exposure of bare soil between crop plants during the season, or in open fields between cropping seasons, creates "leaks" of nutrients from the system. Modern agriculture has come to rely heavily upon nutrient inputs derived or obtained from petroleum-based sources to replace these losses. Sustainability requires that these leaks be reduced to a minimum and recycling mechanisms be reintroduced and strengthened. Ultimately, human societies need to find ways to return nutrients consumed in agricultural products back to the fields, the agroecosystems that consumed and produced them in the first place.

Population Regulating Mechanisms

Through a complex combination of biotic interactions and limits set by the availability of physical resources, population levels of the various organisms are controlled, and thus eventually link to and determine the productivity of an ecosystem. Selection through time tends toward the establishment of the most complex structure biologically possible within the limits set by the environment, permitting the establishment of diverse trophic interactions and niche diversification. Due to human-directed genetic selection and domestication, as well as the overall simplification of agroecosystems (i.e., the loss of niche diversity and a reduction in trophic interactions), populations of crop plants or animals are rarely self reproducing or self regulating. Human inputs in the form of seed or control agents, often dependent on large energy subsidies, determine population sizes. Biological diversity is reduced, natural pest control systems are disrupted, and many niches or microhabitats are left unoccupied. The danger of catastrophic pest or disease outbreak is high, often despite the availability of intensive human interference and inputs. A focus on sustainability requires the reintroduction of the diverse structures and species relationships that permit the functioning of natural control and regulation mechanisms. We must learn to work with and profit from diversity, rather than focus on agroecosystem simplification.

Dynamic Equilibrium

The species richness or diversity of mature ecosystems permits a degree of resistance to all but very damaging perturbations. In many cases, periodic disturbances ensure the highest diversity, and even highest productivity (Connell, 1978). System stability is not a steady state, but rather a dynamic and highly fluctuating one that permits ecosystem recovery following disturbance. This promotes the establishment of an ecological equilibrium that functions on the basis of sustained resource use which the ecosystem can maintain indefinitely and which can even shift if the environment changes. At the same time, rarely do we witness what might be considered large-scale disease outbreaks in healthy, balanced ecosystems. With a reduction of natural structural and functional diversity, much of the resilience of the system is lost, and constant human-derived external inputs must be maintained. An overemphasis on maximizing harvest outputs upsets the former equilibrium and leads to a dependence on outside interference. To reintegrate sustainability, the emergent qualities of system resistance and resiliency must once again play a determining role in agroecosystem design and management.

We need to be able to analyze both the immediate and future impacts of agroecosystem design and management so we can identify the key areas in each system on which to focus the search for alternatives or solutions to problems. We must learn to be more competent in our agroecological analysis in order to avoid problems or negative changes before they occur, rather than struggling to reverse the problems after they have been created. The agroecological approach provides us one such alternative (Altieri, 1995; Gliessman, 1998).

Applying Agroecology

The process of understanding agroecosystem sustainability has its foundations in two kinds of ecosystems: natural ecosystems and traditional (also known as local or indigenous) agroecosystems. Both provide ample evidence of having passed the test of time in terms of long-term productive ability, but each offers a different knowledge base from which to understand this ability. Natural ecosystems are reference systems for understanding the ecological basis for sustainability in a particular location. Traditional agroecosystems provide many examples of how a culture and its local environment have coevolved with time through processes that balance the needs of people, expressed as ecological, technological, and socioeconomic factors. Agroecology, defined as the application of ecological concepts and principles to the design and management of sustainable agroecosystems (Gliessman, 1998), draws on both to become a research approach that can be applied to converting unsustainable and conventional agroecosystems into sustainable ones.

Natural ecosystems reflect a long period of evolution in the use of local resources and adaptation to local ecological conditions. They have each become complex sets of plants and animals that coinhabit a given environment, and as a result, provide extremely useful information for the design of more locally adapted agroecosystems. As I have suggested (Gliessman, 1998), "the greater the structural and functional similarity of an agroecosystem to the natural ecosystems in its biogeographical region, the greater the likelihood that the agroecosystem will be sustainable." If this suggestion holds true, natural ecosystem structures and functions can be used as benchmarks or threshold values for more sustainable systems. Scientists have begun to explore how an understanding of natural ecosystems can be used to guide our search for sustainable agroecosystems that respect and protect the environment and natural resources (Soule and Piper, 1992; Jackson and Jackson, 2002).

Traditional and indigenous agroecosystems are different from conventional systems in that they developed originally in times or places where inputs other than human labor and local resources were generally not available or desirable to the local people. Production takes place in ways that demonstrate people's concerns about long-term sustainability of the system, rather than solely maximizing output and profit. Traditional systems continue to be important as the primary sources of food production for a large part of the populations of many developing countries, while at the same time maintaining their foundations in ecological knowledge (Wilken, 1988; Altieri, 1990). This reality demonstrates their importance for the development of sustainable agroecosystems. This is especially true today when so many modern conventional agroecosystems have caused severe degradation of their ecological foundations, as socioeconomic factors have become the predominant forces in the food system (Altieri, 1990). Many traditional agroecosystems are actually very sophisticated examples of the application of ecological knowledge, and can serve as the starting point for the conversion to more sustainable agroecosystems in the future. The traditional Mesoamerican intercrop of corn (*Zea mays* L.), bean, and squash is a well-known cropping system where higher yields in the mixtures come about due to a complex of interactions among components of the agroecosystem (Amador and Gliessman, 1990). Examples of such interactions range from the increased presence of beneficial insects due to attractive microclimates and a greater abundance of pollen and nectar sources (Letourneau, 1986), to biologically fixed N being made available to corn through mycorrhizal fungi connections with roots of bean (Bethlenfalvay et al., 1991).

How can agroecology link our understanding of natural ecosystem structure and function with the knowledge inherent in traditional agroecosystems? On the one hand, the knowledge of place that comes from understanding local ecology is an essential foundation. Another is the local experience with farming that has its roots in many generations of living and working within the limits of that place. We put both of these approaches together when we work with farmers going through the transition process to more environmentally sound management practices, and thus realize the potential for contributing to long-term sustainability. This transition is already occurring. Many farmers, despite the heavy economic pressure on agriculture, are in the process of converting their farms to more sustainable design and management (National Research Council, 1989; OAC/SCOAR, 2003). In California the dramatic increase in organic acreage for a range of crops has been based largely on farmer innovation (Swezey and Broome, 2000). It is incumbent that agroecologists play an important role in contributing to this conversion process.

Converting an agroecosystem to a more sustainable design is a complex process. It is not just the adoption of a new practice or a new technology. There are no silver bullets. Instead, this conversion uses the agroecological approach described above. The farm is perceived as part of a larger system of interacting parts, an agroecosystem. We must focus on redesigning that system in order to promote the functioning of an entire range of different ecological processes (Gliessman, 1998). In a study of the conversion of conventional strawberries (*Fragaria* × *Ananassa* Rozier) to organic management, several changes were observed (Gliessman et al., 1996). As the use of synthetic chemical inputs was reduced or eliminated and recycling was emphasized, agroecosystem structure and function changed as well. A range of processes and relationships began to transform, beginning with improvement in basic soil structure, an increase in soil organic matter content, and greater diversity and activity of beneficial soil biota. Major changes began to occur in the activity and relationships among weed, insect, and pathogen populations, and in the functioning of natural control mechanisms. For example, predatory mites gradually replaced the use of synthetic acaracides for the control of two-spotted spider mites (*Tetranychus urticae* Koch), the most common arthropod pest in strawberries in California.

Ultimately, nutrient dynamics and cycling, energy use efficiency, and overall agroecosystem productivity are affected. Changes may be required in day-to-day management of the farm, planning, marketing, and even philosophy. The specific needs of each agroecosystem will vary, but the principles for conversion listed in Table 2–1 can serve as general guidelines for working through the transition. It is the role of the agroecologist to help the farmer measure and monitor these changes during the conversion period in order to guide, adjust, and evaluate the conversion process. Such an approach provides an essential framework for determining the requirements for and indicators of sustainable agroecosystem design and management.

Comparing Ecosystems and Agroecosystems

The key to developing sustainability is building a strong ecological foundation under the agroecosystem, using the ecosystem knowledge inherent to agroecology as discussed above. This foundation then serves as the framework for producing the sustainable harvests needed by humans. In order to maintain sustainable harvests, though, human management is a requirement. Agroecosystems are not self sustaining, but rely on natural processes for maintenance of their productivity. An

Table 2–1. Guiding principles for the process of conversion to sustainable agroecosystems design and management (modified from Gliessman, 1998).

- Shift from through-flow nutrient management to recycling of nutrients, with increased dependence on natural processes, such as biological N fixation and mycorrhizal relationships.
- Use renewable sources of energy instead of nonrenewable sources.
- Eliminate the use of nonrenewable off-farm human inputs that have the potential to harm the environment or the health of farmers, farm workers, or consumers.
- When materials must be added to the system, use naturally occurring materials instead of synthetic, manufactured inputs.
- Manage pests, diseases, and weeds instead of "controlling" them.
- Reestablish the biological relationships that can occur naturally on the farm instead of reducing and simplifying them.
- Make more appropriate matches between cropping patterns and the productive potential and physical limitations of the farm landscape.
- Use a strategy of adapting the biological and genetic potential of agricultural plant and animal species to the ecological conditions of the farm rather than modifying the farm to meet the needs of the crops and animals.
- Value most highly the overall health of the agroecosystem rather than the outcome of a particular crop system or season.
- Emphasize conservation of soil, water, energy, and biological resources.
- Incorporate the idea of long-term sustainability into overall agroecosystem design and management.

agroecosystem's resemblance to natural ecosystems allows the system to be sustained, in spite of the long-term human removal of biomass, without large subsidies of nonrenewable energy and without detrimental effects on the surrounding environment.

Table 2–2 compares natural ecosystems with three types of agroecosystems in terms of several ecological criteria. Traditional agroecosystems most closely resemble natural ecosystems, since they most often are focused on the use of locally

Table 2–2. Emergent properties of natural ecosystems, traditional agroecosystems, conventional agroecosystems, and sustainable agroecosystems. Agroecosystem properties are most applicable to the farm scale and for the short- to medium-term time frame.†

Emergent ecological property	Natural ecosystem	Agroecosystem type		
		Traditional	Conventional	Sustainable
Productivity (process)	medium	medium	low/med.	med./high
Species diversity	high	med./high	low	medium
Structural diversity	high	med./high	low	medium
Functional diversity	high	med./high	low	med./high
Output stability	medium	high	low/med.	high
Biomass accumulation	high	high	low	med./high
Nutrient recycling	high	high	low	high
Tropic relationships	high	high	low	med./high
Natural population regulation	high	high	low	med./high
Resistance	high	high	low	medium
Resilience	high	high	low	medium
Dependence on external human inputs	low	low	high	medium
Autonomy	high	high	low	high
Human displacement of ecological processes	low	low	high	low/med.
Sustainability	high	med./high	low	high

† Modified from Odum (1984), Conway (1985), Altieri (1995), and Gliessman (1998).

available and renewable resources, local use of agricultural products, and the return of biomass to the farming system. Sustainable agroecosystems are very similar in many properties, but they are more dissimilar in others because of the probable focus on export of harvest to distant markets, the need to purchase a significant part of their nutrients externally, and the much stronger impact of market systems on agroecosystem diversity and management. Compared with conventional systems, sustainable agroecosystems have somewhat lower and more variable yields due to the weather variation that occurs from year to year. Such reductions in yields can be more than offset, from the perspective of sustainability, through the advantages gained in reduced dependence on external inputs, more reliance on natural controls of pests, and reduced negative off-farm impacts of farming activities.

FUTURE PERSPECTIVES

Problems in agriculture create the pressures for the changes that will bring about a sustainable agriculture. However, it is one thing to express the need for sustainability, and quite another to actually quantify it and bring about the changes that are required. Designing and managing sustainable agroecosystems, as an approach, is in its formative stages. Initially it builds upon the fields of ecology and agricultural science and is emerging as the science of agroecology. This combination can play an important role in developing the understanding necessary for a transition to sustainable agriculture.

But sustainable agriculture is more. It takes on a cultural perspective as the concept expands to include humans and their impacts on agricultural environments. Agricultural systems are a result of the coevolution that occurs between culture and environment, and a sustainable agriculture values the human as well as the ecological components. Our small pool in the stream becomes the focal point for changing how we do agriculture, but that change must occur in the context of the human societies within which agriculture is practiced, the whole stream in our analogy.

All agricultural systems can no longer be viewed as strictly production activities driven primarily by economic pressures. We need to reestablish an awareness of the strong ecological foundation upon which agriculture originally developed and ultimately depends. Too little importance has been given to the "downstream" effects that are manifest off the farm, either by surrounding natural ecosystems or by human communities. We need an interdisciplinary basis upon which to evaluate these impacts.

In the broader context of sustainability, we must study the environmental background of the agroecosystem, as well as the complex of processes involved in the maintenance of long-term productivity. We must first establish the ecological basis of sustainability in terms of resource use and conservation, including soil, water, genetic resources, and air quality. Then we must examine the interactions among the many organisms of the agroecosystem, beginning with interactions at the individual species level and culminating at the ecosystem level as our understanding of the dynamics of the entire system is revealed.

Our understanding of ecosystem-level processes should then integrate the multiple aspects of the social, economic, and political systems within which agroecosystems function, making them even more complex systems. Such an integration of ecosystem and social system knowledge about agricultural processes will not only lead to a reduction in synthetic inputs used for maintaining productivity; it will also permit the evaluation of such qualities of agroecosystems as the long-term effects of different input–output strategies, the importance of the environmental services provided by agricultural landscapes, and the relationship between economic and ecological components of sustainable agroecosystem management. By properly selecting and understanding the "upstream" inputs into agriculture, we can be assured that what we send "downstream" will promote a sustainable future.

ACKNOWLEDGMENTS

The author is extremely grateful for support provided by the Alfred Heller Endowed Chair for Agroecology. An early version of this chapter was rigorously discussed in a Kellogg Foundation National Fellowship Forum, and the valuable input from the Fellows in Group VI is warmly acknowledged. Detailed edits from Diane Rickerl, Chuck Francis, and an anonymous reviewer are also much appreciated. Joji Muramoto graciously assisted with the formatting of Fig. 2–1.

STUDY QUESTIONS

1. What kinds of changes need to be made in the design and management of agroecosystems so that we can come closer to farming in "nature's image"?

2. Describe a characteristic or component of a traditional farming system that would find widespread application in conventional farming systems if sustainability were a primary goal.

3. What are some of the primary key variables that determine the length of time necessary for converting a farm from nonsustainable to sustainable management?

4. What do you feel are the key differences in how nutrient cycles operate in natural ecosystems as compared with conventional agroecosystems? How would you take these differences into account in developing a more sustainable design and management strategy for agriculture?

5. What is meant by "redesigning" an agroecosystem in the process of conversion to more sustainable agroecosystem research and development? Give an example of a change you would incorporate into a cropping system of your choice that demonstrates your knowledge of the concept.

REFERENCES

Altieri, M.A. 1990. Why study traditional agriculture? p. 551–564. *In* C.R. Carroll et al. (ed.) Agroecology. McGraw-Hill, New York.

Altieri, M.A. 1995. Agroecology: The scientific basis of alternative agriculture. 2nd ed. Westview Press, Boulder, CO.

Amador, M.F., and S.R. Gliessman. 1990. An ecological approach to reducing external inputs through the use of intercropping. p. 146–159. *In* S.R. Gliessman (ed.) Agroecology: Researching the ecological basis for sustainable agriculture. Springer-Verlag, New York.

Bethlenfalvay, G.J., M.G. Reyes-Solis, S.B. Camel, and R. Ferrera-Cerrato. 1991. Nutrient transfer between the root zones of soybean and maize plants connected by a common mycorrhizal inoculum. Physiol. Plant. 82:423–432.

Borman, F.H., and G.E. Likens. 1967. Nutrient cycles. Science. 155:424–429.

Connell, J.H. 1978. Diversity in tropical rain forests and coral reefs. Science 199:1302–1310.

Conway, G.R. 1985. Agroecosystem analysis. Agric. Admin. 20:31–55.

Flora, C. (ed.) 2001. Interactions between agroecosystems and rural communities. Adv. in Agroecology. CRC Press, Boca Raton, FL.

Francis, C., G. Lieblein, S. Gliessman, T.A. Breland, N. Creamer, R. Harwood, L. Salomonsson, J. Helenius, D. Rickerl, R. Salvador, M. Wiendehoeft, S. Simmons, P. Allen, M. Altieri, J. Porter, C. Flora, and R. Poincelot. 2003. Agroecology: The ecology of food systems. J. Sustain. Agric. 22(3):99–119.

Gliessman, S.R. 1998. Agroecology: Ecological processes in sustainable agriculture. Lewis/CRC Press, Boca Raton, FL.

Gliessman, S.R. (ed.) 2001. Agroecosystem sustainability: Toward practical strategies. Adv. in Agroecology. CRC Press, Boca Raton, FL.

Gliessman, S.R., M.R. Werner, S. Swezey, E. Caswell, J. Cochran, and F. Rosado-May. 1996. Conversion to organic strawberry management changes ecological processes. Calif. Agric. 50:24–31.

Jackson, D.L., and L.L. Jackson. 2002. The farm as natural habitat. Island Press, Washington, DC.

Kimbrell, A. (ed.) 2002. Fatal harvest: The tragedy of industrial agriculture. Island Press, Washington, DC.

Letourneau, D.K. 1986. Associational resistance in squash monoculture and polycultures in tropical Mexico. Environ. Entomol. 15:285–292.

National Research Council. 1989. Alternative agriculture. National Acad. Press, Washington, DC.

Odum, E.P. 1971. Fundamentals of ecology. W.B. Saunders, Philadelphia, PA.

Odum, E.P. 1984. Properties of agroecosystems. p. 5–12. *In* R. Lowrance et al. (ed.) Agricultural ecosystems: Unifying concepts. John Wiley & Sons, New York.

Odum, E.P. 1996. Ecology: Bridging science and society. Sinauer Associates Inc., Sunderland, MA.

Organic Agriculture Consortium (OAC)/Scientific Congress on Organic Agriculture Research (SCOAR). 2003. Organic Ag Info [Online.] Available at www.organicaginfo.org (verified 1 Oct. 2003). Econ. Res. Serv. Issues Center. Washington, DC.

Pimentel, D., and M. Pimentel (ed.) 1997. Food, energy, and society. 2nd ed. Univ. Press of Colorado, Niwot.

Rappaport, R.A. 1971. The flow of energy in an agricultural society. Sci. Am. 224:117–132.

Soule, J.D., and J.K. Piper. 1992. Farming in nature's image. Island Press, Washington, DC.

Swezey, S.L., and J. Broome. 2000. Growth predicted in biologically integrated and organic farming. Calif. Agric. 54:26–35.

Tivy, J. 1990. Agricultural ecology. Longman Scientific and Technical, London.

Wilken, G.C. 1988. Good farmers: Traditional agricultural resource management in Mexico and Central America. Univ. California Press, Berkeley.

3 Soil Dynamics, Plant Nutrition, and Soil Quality

CHARLES FRANCIS

University of Nebraska
Lincoln, Nebraska

Environmental determinism suggests that human societies have prospered or have been hungry because of their physical locations and the soil and water resources available and sustainable with time (Diamond, 1997). Jared Diamond's widely read *Guns, Germs, and Steel,* which won the Pulitzer Prize, eloquently described the rise and fall of cultures during the past 10 000 years. The grasslands that supported populations of grazing herbivores and early human culture in what is now the Sahara Desert changed drastically as a result of overgrazing and climate change. The fertile soils, grand irrigation schemes, and diverse cultivation of cereals, fruits, and vegetables in Southwest Asia sustained thriving cultures until soil salinity reduced crop yields and human populations exceeded the carrying capacity of the land (Hillel, 1992). Hillel described how dominant cultures pushed out their boundaries to exploit soil fertility elsewhere, until their supply lines were overextended or until those lands too became infertile. People who lived on trade routes could survive by barter or sale of products on the move, essentially tapping into the soil fertility in distant places. The message that comes through the centuries of human history is that fertile soils sustain human life and a complex ecosystem. To ignore such lessons is to go blindly toward the future.

In an insightful overview of the importance of soils to humankind, scientist Dan Yaalov (2000) paraphrased Leonardo da Vinci: "Why do we know more about distant celestial objects than we do about the ground below our feet?" Yaalov went on to describe the study and appreciation of the intricacies of soil development as a complex result of parent substrate, climate, and vegetation. He identified language barriers and slow communications as the reasons that important research in Russia was not rapidly available or even accepted in the West. From the perspective of temporal and spatial scales, we could interpret this separation in distance and language, even social and cultural isolationism, as preventing the sharing of information and meaning across national boundaries.

As scientists began to study soils more in depth, they added topography, time, and soil biology to the list of causative factors in soil genesis. Most recently, the complex biological webs that characterize interactions of belowground populations of micro- and macroorganisms have been recognized as critical to soil fertility and quality (Doran and Jones, 1996). As many authors now point out, our management of this fragile soil ecosystem will ultimately determine our own survival as a species.

A hierarchy of spatial scale can be useful to summarize our research and understanding of the soil system that sustains crop and animal production. Careful attention to the myriad life forms that inhabit the soil is relatively new among soils research areas. Understanding has progressed through species discovery and taxonomy, through identifying functions of individual species, to a focus on the complexity of interactions among these millions of organisms that inhabit a teaspoon of soil. How cultivation, fertilization, and crop sequences in the field affect soil organic matter and the complex web of life underground is central to our successful management of cropping and crop–animal agroecosystems.

Most research on soil fertility has focused on plant nutrition and the search for optimum fertilizer rates to reach maximum crop or pasture production and profits. Likewise, in organic production systems, emphasis has been on how to maintain adequate soil nutrient levels without application of synthetic fertilizers. Whole-farm nutrient budget calculations and attention to balance cropping fields and animal numbers in integrated systems that are consistent with the farm's internal resources have been important contributions of researchers and organic producers to all of agriculture. Interest at the landscape, watershed, and regional levels has been more toward water quality measurements in streams leaving farms and watersheds. A major step and incentive for understanding regional systems has been the identification of nonpoint source entry of nutrients and chemical pesticides from farmlands into tributaries of the Mississippi River, with the consequence of pollution in a large area (more than 20 000 km^2 [8000 sq. miles]) called the Dead Zone in the Caribbean Sea.

This chapter reviews and summarizes research information from each of several levels in the spatial scale, with emphasis on how this information can help us design soil management strategies for sustained, long-term production. The objective is not to provide a comprehensive review of recent research, as this was well addressed by Magdoff et al. (1997), who described nutrient cycling and nutrient transformations across a scale of spatial hierarchy from plant to field to farm to watershed and global levels. The purpose here is rather to illustrate how soil-related factors at each level of scale are complex, and how they relate to diverse factors at that level and others. Design of a soil fertility strategy is a specific application of the principles shown in Fig. 10–2 and 10–4 of this book (Chapter 10, Francis et al., 2004), where biophysical factors at any level interact with many other types of factors (e.g., climate, economics, crop subsidies) across the spatial scale. Many complex processes and factors contribute to soil quality, and understanding nutrient dynamics is essential to designing an efficient soil fertility program (Doran and Jones, 1996).

The importance of a systems focus that goes beyond the narrow confines of single disciplines is emphasized in each chapter section. Such a focus will help us heed the admonition of Doran and Sims (2002) that "soils and the environment...are more and more besieged by sophisticated technologies that too often pay little attention to all but their own sphere of influence." Doran and Sims undoubtedly refer as well to the scientists who develop these technologies. We are well advised to approach the challenge of sustaining soil fertility in a holistic manner, even taking nature and natural ecosystems as a guide in our quest (Soule and Piper, 1992).

IMPORTANCE OF SCALE AND HIERARCHY

How we view soils and view the world depends on where we stand, what accumulated experiences we bring to that place and moment, and how wide is our field of vision. Different examples of art from the Midwest were invoked in Chapter 1 (Rickerl and Francis, 2004) to illustrate our capacities for multidimensional thinking, with a progression of Grant Wood paintings (Fig. 1–1, 1–2, 1–3) from unidimensional, to two- and three-dimensional perspectives used as an analogy for how we view systems. We may not explicitly recognize these differences in dimensionality until they are pointed out. Most people are unimpressed with Picasso's paintings of figures with discontinuous body parts and strange configurations that defy common understanding and our normal perception of what a body should look like. Only when we realize that his figures may represent the same body when viewed from different angles does the painting take on meaning, and his genius become revealed. Likewise, we each look at the agroecosystem from our own discipline, and add meaning to that view through our individual experiences.

One way to view the complexity of issues involved in understanding soil fertility and the design of crop and crop–animal systems to supply nutrients for successfully maintaining production is through a scale of spatial hierarchy. This is illustrated in Fig. 3–1, with levels in scale from soil microorganisms, through individual organisms to fields, whole farms, watersheds, and regions. Examples of specific factors that influence nutrients and fertility at the same and at different levels are listed in Table 3–1. Problems in production systems can be identified in terms of where they can best be studied in this hierarchy, and it is essential to know where it is appropriate to apply the results. If we recognize a nutrient deficiency in a maize (*Zea mays* L.) field, careful observation is needed to determine how generally that deficiency occurs in the field and the farm, and such observations are made at different orders of scale. We may also learn from other factors such as field crop or grazing history or previous cultural practices why there is a deficiency in the current crop. Such history adds new meaning and value to the observation. In looking more narrowly at the situation, we may discover a microscale problem in that place due to a sand lens or specific soil biology condition. We can look for mechanisms by going to a finer level of scale to study the situation. Thus our level of inquiry moves up and down the scale.

Thoughtful planning for which level of the scale hierarchy is most appropriate to study a given problem is important for the design of research and application of results. We must be realistic in the interpretation of results and not apply recommendations beyond the appropriate level of scale or they may become meaningless. An observed nutrient deficiency is best studied in the areas of the field or in entire fields where that symptom occurs. We can add credibility to results by setting up similar trials in other fields that exhibit similar symptoms or by conducting the experiment over several cropping seasons. If we are able to discover the nature of the problem and also a reasonable solution, such as through application of a fertilizer or green manure crop or change in management, we can logically use that solution in those fields where the situation is similar, even beyond the fields where experiments were conducted. We must be cautious, however, not to extrapolate the results and solutions to far distant fields where symptoms may be the same but soil conditions or other factors are so different that the solution is ineffective or economically not feasible. For this reason, Magdoff et al. (1997) chose to review

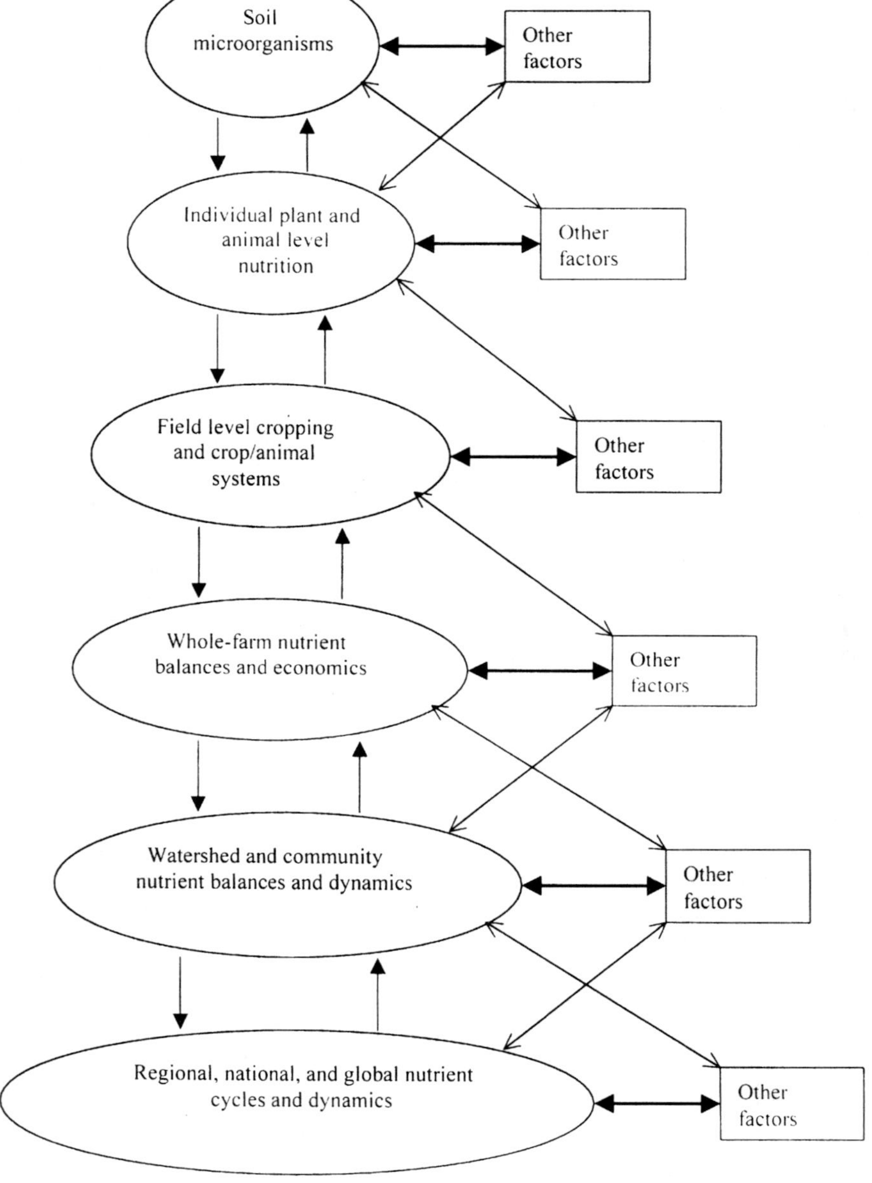

Fig. 3–1. Interactions and processes among soil-related organisms, plants, and animals across different levels in a spatial hierarchy.

Table 3–1. Factors that influence soil dynamics and plant nutrition at different levels in the spatial hierarchy.

Spatial level in hierarchy	Factors at same level	Factors at lower level	Factors at higher level
Soil microorganisms and macroorganisms	Diversity of species, soil organic matter	Soil chemistry, soil moisture content	Tillage, crop species, animal grazing
Individual plant and animal nutrition	Species mixture, weed populations	Soil fertility, soil biology	Tillage, fertilizer applications
Field cropping and crop–animal systems	Crop rotations, animals' grazing residues	Crop and animal genotypes	Timing of field operations
Whole-farm nutrient balances, economics	Labor, management, enterprise balance	Crop rotations, crop–animal numbers	Local and federal regulations, prices
Watershed, community nutrient balances, dynamics	Human and livestock population, difficulty in measurements	Farm size, applied nutrients, and farm balance	Federal regulations, global prices for fossil fuels, commodities
Regional, national, global nutrient cycles and dynamics	International accords, knowledge about global impacts	Aggregation of all above factors	not applicable

nutrient cycling at the soil–plant interface, the field and farm levels, and broader issues at the watershed, regional, and global levels. Their excellent review is the source of a number of citations used in this chapter. Magdoff and colleagues also recognized the importance of time. Biological and other processes happen more rapidly at lower levels in the spatial hierarchy (e.g., nutrient chemical transformations in the soil solution or multiplication of soil bacterial populations) than at higher levels (e.g., changes in soil organic matter content at the farm level or formation of a dead zone in the Caribbean Sea due to excess nutrient loading). Understanding how to focus research can inform our study of agroecosystems, and specifically soil fertility, and guide our application of results in an efficient, appropriate, and effective manner.

SOIL MICROORGANISM AND MACROORGANISM LEVEL

Soil scientists report that more than 100 000 different species of soil microorganisms exist, although only a small portion (perhaps <10%) has been identified (Kennedy and Doran, 2002). Thousands more species of macrofauna inhabit the upper strata of productive soils. Soil macroorganisms include a diverse array of soil fauna such as nematodes, beetles, flies, ants and termites, bees and wasps, springtails, mites, flatworms and earthworms, centipedes and millipedes, as well as microorganisms including fungi, bacteria, and actinomycetes. Many algal species live in the soil. Even some mammals, birds, and reptiles spend much of their lives underground. Operating at different trophic levels in the soil ecosystem, these soil organisms (except for the soil algae) depend ultimately on plant photosynthesis for their energy (Lampkin, 1990). Crop residues, including all parts not harvested and removed from the field plus weeds, provide a major source of soil organic matter (whole plant and field levels) that highly influences microorganism activity. Residues are part of the active fraction of soil organic matter that is broken down by decomposers, and their residues are in turn consumed and broken down by organisms at a higher trophic level. Microorganism activity is especially high near roots in the rhizosphere (within 1–2 mm from root surfaces) where old cells are available along with root exudates, making the rhizosphere a vital area for nutrient exchange and uptake by plants. Carbon, N, and other elements move up through the food chain, and soil organisms are subject to a wide range of control measures at the soil level, including populations of different species, available nutrients, and microclimatic factors (Magdoff and van Es, 2000).

Soil organism diversity is seen as a major key to soil's capacity to buffer extreme chemical changes (Kennedy and Doran, 2002), although there is limited experimental evidence to strongly support this conclusion. Factors that influence the level of soil organism diversity include amount of soil organic matter, speed of nutrient cycling (a function of temperature and soil moisture plus type of crop residue management), level of plant growth (a function of crop choice, weeds, rotations, and amount of legumes in mixture), soil type, and soil structure, among others. The tremendous stability of soil structure, composition, chemical reactions, and function is remarkable, in spite of different cultural practices, applications of fertilizers and pesticides, and other management practices. Soil pH, organic matter content, and structure change as a result of human decisions, yet the soil mass is amazingly resilient due to the complex and diverse soil biology found in most soil types.

Although many of the internal controls of microorganism populations operate within the microenvironment where they live, most processes are highly influenced by factors at other levels in the spatial hierarchy (e.g., regional storm fronts and large rainfall events, global energy price hikes that discourage N fertilizer applications). The amount of crop residue returned to the soil depends greatly on the crop and how much residue is removed from the field. Maize harvested for grain with a yield of 10 Mg ha^{-1} will return about 10 Mg ha^{-1} of dry matter in stalk, leaf, husk, and cob residue, plus that in the root system. A similar maize crop harvested for silage may have <1 Mg ha^{-1} returned as residue plus that in the root system. A soybean [*Glycine max* (L.) Merr.] crop that yields 3 Mg ha^{-1} seed will return 2 to 3 Mg ha^{-1} of crop residues plus those in the roots. Thus, crop choice and destination at harvest, both farm-level decisions, strongly influence soil organic matter and micro- and macroorganism activity.

Symbiotic microorganism activity is highly important in agroecosystems, especially the fixation of N by *Rhizobium* spp. and *Bradyrhizobium* spp. bacteria and P mobilization by various mychorrizal fungus species. Nitrogen-fixing bacteria in conjunction with legumes are responsible for the largest fraction of naturally available N for current and subsequent crop growth, fixing 40 to more than 100 kg N ha^{-1} yr^{-1}. But rhizobial activity is highly influenced by soil N status, and when a large quantity of N is supplied through animal manure or chemical fertilizer sources and/or mineralization, the fixation is drastically reduced. Likewise, the level of mychorrizal fungus activity is highly influenced by soil P levels. Application of fertilizer P can greatly reduce mychorrizal fungal action and the amount of P mobilized by the process. Thus, management decisions on fertilizer application can greatly influence soil microbial activity.

Natural climatic factors also strongly influence soil organic matter and biological activity. Temperature and rainfall not only impact growth and reproduction of soil organisms, they influence crop dry matter production and the amount of residue returned to replenish nutrients in the soil organic matter. Fine-textured soils such as clays and clay loams usually have greater amounts of soil organic matter than sandy or course-textured soils. Organic matter is less rapidly broken down in clay soils if aeration is less, and the finer soil particles tend to protect the organic fraction. Topographic position in the field also influences drainage, soil moisture, and thus the rate of soil organism activity and residue decomposition. These natural factors are all influenced by management decisions, including manner and depth of tillage, timing of field preparation, crop rotation, applications of chemical fertilizers and animal manures, and use of green manure crops. These activities and decisions all occur at other levels in the spatial hierarchy.

Economic and policy factors from community to regional to national level can influence soil organisms and organic matter. Relative prices of crops will often determine what is planted, and local crop prices will in large part be a function of global commodity prices. Decisions to set aside land and provide federal incentive payments influence land use and tillage for a decade or more. The recent removal of base acre requirements for federal crop supports in the USA removed some constraints on crop choice and how many acres could be planted of each, which indirectly but significantly impacted amounts of residue returned to the soil as the crop mix changed on a farm. New interest in C payments is impacting management decisions on how much residue to preserve on the soil surface, thus sequestering that C for a longer period of time and changing soil biology. Economic and policy factors all operate at levels far above the soil microenvironment, and they can all potentially impact soil micro- and macroorganisms.

INDIVIDUAL PLANT- AND ANIMAL-LEVEL NUTRITION

Plant growth and crop yields are strongly influenced by soil nutrient status, and N and other macroelements are often the most limiting factors that determine soil productivity. Our understanding of the natural processes of nutrient cycling in soils is hampered by the high levels of application of soluble chemical fertilizers in most agroecosystems. Although it is possible to learn much about N, P, and other cycles in the soil and wider environment from studies of natural prairies and other biomes, the absence of human intervention and management of these areas create differences that make it difficult to extrapolate to cultivated agroecosystems. This substantiates the importance of research in organic production systems and work on systems that depend entirely on green manures and rotations for their soil nutrient supply (see Karlen et al., 1994; Lotter, 2003).

Animal nutrition in pasture production systems is highly dependent on quality, consistency, and year-long availability of feed supply. Nutrient cycling is accelerated by including grazing animals in the system, since forage is consumed and more soluble nutrients excreted with urine and manure in a dispersed pattern across the field. Health and productivity of individual animals are governed by their feed supply and the quality of that feed. Mixtures of grasses and legumes in pastures provide a better diet than monocultures. In fact, pasture diets of only young alfalfa (*Medicago sativa* L.) or clover (*Trifolium* spp. and others) can be dangerous due to lack of fiber and potential for causing bloat. Often, temporal diversity in feed supply is introduced on a daily basis by feeding hay early and then putting animals out into high-quality legume pastures.

The nutrition available to individual plants and animals is determined strongly by the management methods employed, including applied fertilizers, crop rotations, dates of planting, interseeded species, as well as soil properties and local weather conditions, especially rainfall and temperature. Early planting into cold soils in northern regions, a management strategy to try and maximize the number of days in the growing season, can result in lack of P available for early plant germination and growth. How a given crop is managed depends on farm size, equipment, and the timeliness of each cultural practice. A decision to plant soybean in one field in early June in Nebraska precludes the use of the same tractor by the same farmer to cultivate a corn field planted a month earlier that badly needs the weeds suppressed. These are interactions at a higher level that impact the nutrients available and competition for resources for the individual plant. On the animal management side, the pressure of work at planting and harvest times on a mixed farm make it difficult to adequately manage cattle at the same time, and individual animal disease or nutrition problems may go undetected for some days. Such interactions represent the trade-offs that occur in a complex farming operation, where labor is limited and the farmer must juggle multiple priorities on a daily or even hourly basis.

FIELD-LEVEL CROPPING AND CROP–ANIMAL SYSTEMS

The field level is where management decisions are most often made and where the impacts of these decisions are most readily observed; it is also the spatial level at which most agronomic and animal nutrition research work has been done. Thus, we have a major store of knowledge about crop yield responses to fertilizer rates, formulations, and dates of application, as well as recommendations for farmers in most areas and for different soils. Likewise, there is ample data about the productivity of different pasture mixtures, stocking rates for livestock, value of stalk grazing after maize harvest, and similar results from enterprise-level research. Most of these recommendations are made in the context of single factor experiments designed to answer questions that assume factor *X* is the principal limitation to production. Factors from other spatial levels will influence how closely the use of recommendations will hold true in a given context. The response of different crop genotypes to fertilizer application, the mixture of species in a grass–clover pasture, and the presence or absence of inoculum for rhizobial development on a legume are examples of factors that may drastically change plant growth and development and the response to recommended fertilizer rates. Factors from higher spatial levels include decisions to graze crop residues during winter, farm-level decisions to do fall vs. spring plowing on the whole or part of the farm, or prices of fertilizer nutrients that depend on global oil prices.

Crop rotations are among the most important practices recommended at the field level for improving yields, reducing pest problems, and lowering the need for applied fertilizers (for a comprehensive review see Karlen et al., 1994). Maize yields are increased from 5 to 20% by rotation with soybean, with an average expected increase of about 10% (see Crookston et al., 1991 and many other reports). Similar yield increases have been reported for soybean and grain sorghum [*Sorghum bicolor* (L.) Moench] in a number of studies in the Midwest. Precisely how effective a given rotation will be in a specific field depends on factors such as the previous year's crop and yield, rainfall up to time of planting and through the season, and whether manure is added to the field and considered in the crop nutrient budget. Also important are factors that influence whether a farmer sticks with a given rotation in the field or not, including changing rules on federal crop support payments, increasing fossil fuel and N prices that may encourage a shift from a high N requirement cereal to a grain legume for one season, or an exorbitantly high price for one crop that may entice people to overplant the area in that one crop.

Fertilizer application rates that are recommended for specific soils and field locations are currently being integrated into models and programs for application in a more site-specific way, using the capabilities of global positioning systems (GPS) mounted on equipment for precisely locating unique areas of each field (see Chapter 9, Caldwell, 2004). Such technology makes possible variable rate applications of fertilizers to match the yield potentials of different parts of the field or to correct specific problems that are not extensive across whole fields. Although site-specific technology is expensive, some farmers are convinced that this pays off in more precise application and not wasting fertilizer resources in areas of the field with low yield potential. The system must be calibrated, however, on the average response under experimental conditions over years of precise experiments, and recommendations generally do not take into account previous year yields, current soil tests, rainfall up to date of planting, or response of a new hybrid to that field. Site-specific technology has significant limitations related to precision of the data used to set up recommendations and applicability to a given field in a given year under a given set of conditions, especially unpredictable rainfall. Nutrient status and crop response at the field level depend on factors from different levels in the spatial or temporal hierarchy.

WHOLE-FARM NUTRIENT BALANCES AND ECONOMICS

Literature on whole-farm systems is far less prominent than what has been published at the individual crop or field level, but there is a growing concern about how nutrient balances can be calculated on the farm and how these can be altered by choice of crops, decisions on systems, and design of crop–animal integrated enterprises. Magdoff et al. (1997) discussed the factors that influence farm-level nutrient cycles and flows, including how well crops and livestock are balanced and integrated. Animal manure may be well dispersed by grazing livestock or applied especially where needed, or in contrast, it may be applied on sites near the barn or feedlot for convenience. Where manure is applied is a function of equipment size and distance to fields (farm level), labor available (farm or community level), or regulations on maximum allowable application rates (watershed, region, or national level). Distribution of nutrients to where they are needed greatly determines the economics of nutrient cycling within the farm. Explicit calculation of nutrient benefits and costs of application by field can provide a good knowledge base upon which to make management decisions.

An accounting of all nutrients that enter and leave the farm is useful in understanding the whole-farm nutrient balance. Nutrients come to the farm as fertilizers and purchased animal manure, lime or other soil additives, purchased feed, N fixed from the atmosphere, and in rain, snowfall, and irrigation water (Magdoff et al., 1997). Nutrients leave the farm as sold products or through erosion, leaching, and volatilization.

For N, Cassman et al. (2002) estimated on a global basis that N fertilizer represents 46% of all N inputs to crop plant nutrition, while 20% is from biological fixation, 12% from atmospheric deposits, 11% from animal manure, 7% from crop residues, and the rest from internal sources. All fertilizer is considered imported into the farm, all crop residues internal to the farm, and others such as manure could be one or the other or a combination of the two. Whether biological N fixation and atmospheric deposits are considered external or internal sources depends on how we define the farm boundaries. How N is used within the farm depends on fertilizer recovery rate, rainfall, soil properties, temperature, and crop yields. Cassman et al. (2002) estimate that fertilizer recovery efficiency in Asian rice production is about

31%, while for maize in the USA it is about 37%. They report large genetic variation in major cereals in grain N accumulation, partially explaining the differences in reported values of N recovery. At the farm level, Cassman et al. (2002) reported greater efficiency in N recovery by maize in more stress-tolerant hybrids, fields with higher yields, conservation tillage management, and higher plant populations. These are all factors at the individual plant and field levels. Improved efficiency may be achieved by spring application, split applications, and careful field and farm N budgeting.

Studies of conventional vs. organic (in Europe often called ecological) systems have revealed the basics of nutrient cycles and flows unconfounded by application of large amounts of chemical fertilizer. Granstedt and Westberg (1993) studied nutrient cycles and budgets at the farm, community, and national levels in Sweden. They found that losses of N and other nutrients from farms were increased when animal density was high and when there was substantial use of off-farm purchased feed. They further observed that N use was greatest in areas with stockless farms. On the basis of a 21-yr study of organic and conventional farming systems, Mäder et al. (2002) reported that crop yields were 20% lower in the organic system, but use of synthetic fertilizer was reduced by 34% and pesticides were reduced by 97% in the organic and integrated farming alternatives. Greater biodiversity on the organic farms accounted for their productivity and ability to cycle nutrients and prevent most pest losses. The success of organic farming practices and whole-farm systems is related to numerous factors at both lower and higher spatial levels. At the soil level a newly converted organic system may not yet be at a biological equilibrium that supports adequate crop plant nutrition. At plant level there may be differences in crop genetic efficiency in using available nutrients or in rapid germination and growth by the crop that will help it compete with weeds for nutrients, water, and light. At the field level the longer crop rotations often found in organic systems may increase soil organic matter, aid in pest control, and enhance soil fertility (Lotter, 2003). At levels beyond the farm, subsidies for specific crops or livestock or for converting the farm may greatly influence the crop and livestock systems decisions and subsequent nutrient flows and cycles on and off the farm.

WATERSHED AND COMMUNITY NUTRIENT BALANCES AND DYNAMICS

Cycles and flows of nutrients at the watershed level generally reflect similar processes at the whole-farm level, but manifested at a larger scale. Moving to a higher level of analysis introduces greater error due to greater diversity within the land area and complexity of crops, animal enterprises, and management combinations. The fates of specific inputs in a watershed are more difficult to determine when compared with a lower level of analysis (Magdoff et al., 1997). Fertilizer N again represents the largest input in watersheds dedicated to agroecosystems, and this is logical as an aggregate of the applications of nutrients in the component fields and farms in the area. Depending on diversity of land use, there may be more or less of the applied nutrients retained in or leaving the watershed. Not surprisingly, nutrients are more readily and frequently lost from intensively cropped watersheds than those that are forested or in pasture or prairie (Correll, 1983). Many watersheds that include major agroecosystems contain a mixture of intensely cultivated fields and some natural areas, and their cycles and losses are intermediate between the two extremes cited here. How much of the land is intensively cultivated will be a function of soil quality and potential crop production (soil and field levels), rainfall and farm size (farm and region levels), proximity to markets and transportation (watershed and region levels), federal farm program subsidies in a given year (national level), and international grain prices (global level).

Much of the research on water quality has been done at the watershed and higher levels of scale, where measurements of dissolved nutrients have been measured in streams and rivers to determine where nonpoint source pollution may originate. Magdoff et al. (1997) pointed out that "the appropriate level for observation and for responsibility remains a question," since it is difficult to precisely trace sources of nutrients and therefore to design a strategy for incentives to reduce their entry into water courses. Nutrient pulses (shorter-term high levels) during the year, especially in the spring after heavy fertilizer applications, accelerate or spread movement of nutrients through the soil profile and into the groundwater. Differences in boundaries among legal property lines, natural divisions between watersheds, and underground water flows make this a highly complicated process to understand and quantify. Any attempt to aggregate data from farms in a watershed and to determine nutrient balances at this level is confounded by the above factors. The same factors listed above for lower and higher levels in the spatial hierarchy would impact decisions in each farm and thus influence the performance of the watershed.

REGIONAL, NATIONAL, AND GLOBAL NUTRIENT CYCLES AND DYNAMICS

A recent issue of the journal *Ambio* (Galloway et al., 2002) included a number of papers on reactive N (active N in the soil solution and organic matter) that explored the importance of this fraction in regional and global agroecosystems and the environment. They summarized a range of positive and negative effects of reactive N:

- Most of the world's population is sustained by food produced with reactive N created by human actions.
- All reactive N created during food and energy production eventually is released to the environment.
- This reactive N is accumulating in the atmosphere, soils, forests, and waters.
- There is a long cascade of detrimental effects, including smog, haze, particles in the air, acid deposits, forest decline, surface water deterioration, coastal eutrification, ozone depletion, and subsequent greenhouse effects.

These regional and global observations are the obvious consequence of human activities that influence N production and application at all the lower levels of the spatial hierarchy, and the accumulating problems often reflect a substantial time during which N movement has occurred. All the factors that have been described in the previous sections have contributed to both the increases in food production made possible by these N applications as well as the considerable negative impacts on the environment and ultimately on human quality of life. Through careful examination of cycles and flows at each level in the hierarchy, we can find alternatives and apply solutions to solve the global challenges listed above.

Nitrogen has been the element of most concern to national research programs, as costs are high for production of this essential element, and its solubility leads to high risk for loss from overapplication of fertilizer to production fields or excessive animal manure or green manure in the system. In Norway, Bleken and Bakken (1997) calculated the amount of N that is applied to production fields, the largest N flow in society, and the impacts this has had on the total N cycle. They estimated that only about 10% of N applied in the primary plant production process reaches the consumer in food, and that N costs depend highly on the prevalent diets in society. Nitrogen cost is defined as the ratio between fertilizer and manure N input and the N in food products (Bleken and Bakken, 1997). This cost ratio is 3 for wheat, 14 for dairy products, and 21 for meat. Bleken and Bakken (1997) concluded that research, education, and policy must focus at the highest trophic level possible, such as finding ways to encourage a more vegetarian diet for example, than at lower trophic levels, such as making compost on the farm. Involvement at all levels is important, of course, but it is apparent from their data that greater impact from scarce resources in research and development would be realized by changing dietary habits. Bleken and Bakken (1997) cited similar national studies on nutrient use and loss in agriculture at farm and larger spatial levels that have been published for Denmark (Hansen, 1989; Sibbeson, 1990), Sweden (Granstedt and Westberg, 1993), the Netherlands (Olsthoorn, 1992), Poland (Sapek and Sapek, 1993), and for the European Union (Brouwer et al., 1995; Isermann, 1991).

There is no way in a few years that we can minimize the challenges of feeding a rapidly growing human population, although all possible means through education and incentives should be used in the longer term to establish a rational level of human population that is more in balance with the earth's carrying capacity. Galloway et al. (2002) outlined a series of opportunities that can be exploited to optimize N management in food, fiber, and energy production as well as reduce the negative environmental impacts of this process. These changes all involve reducing the amount of N that is converted to reactive forms by the human population:

- Increase N-use efficiency in the food production system.
- Recycle reactive N within agroecosystems and forests.
- Decrease N emissions from fossil fuel combustion by capture or eliminating formation.
- Convert reactive N back to nonreactive N before it is lost to the environment.

The editors of the special edition of *Ambio* (Galloway et al., 2002) refer to the 2nd International Nitrogen Conference held in Maryland in 2001 where a series of 35 recommendations were presented for research, education, and management to help solve the problem of excessive reactive N in the ecosystem (Galloway et al., 2001). The recommendations were based on the recognition that human food and energy production are drastically changing the earth's N cycle, and that although increased food production is essential, many of these changes are not favorable to humans and other species. The conference participants urged creativity and commitment for countries to make better N management choices in food production, consumption, and nutrient cycling, as well as focus concern on environmental protection.

Results from these two global conferences provide compelling evidence that we must carefully study the earth's N cycles and our impacts on this element at every level in the spatial hierarchy. Although N is one of the most easily measurable nutrients in the system, we should be equally concerned about other elements that are widely applied in agroecosystems and that also cause negative environmental consequences. Phosphorus loss from agricultural fields and its accumulation in lakes and other water bodies is a prime example. High P levels cause water contamination and changes in water chemistry that lead to growth of unwanted populations of algae and plants that drastically change ecosystem functions. These challenges are all related to management decisions on individual farms, yet the economic and political decisions at several higher levels determine what economic decisions are rational for each farmer. Organic agriculture is certainly not exempt from the same economic influences, but many of the methods used by organic farmers can and do help to resolve the challenges described above.

CONCLUSIONS

The dynamic processes that influence soil biology and chemistry, plant nutrition, and soil quality are complex, and they operate through time at different levels of spatial hierarchy. This chapter lists a number of factors that are important at each spatial level and how they are impacted by other factors at the same spatial level as well as by factors at lower and higher spatial levels. Several relevant reviews have provided excellent windows on the literature in nutrient cycling (Magdoff et al., 1997), impacts of crop rotations (Karlen et al., 1994), and organic farming (Lotter, 2003) that reveal much about our understanding of how these factors interact. Specific influences across the spatial scale from soil microorganism populations to global energy prices can significantly impact crop and pasture plant nutrition.

Most research and thus most detailed understanding of complex soil systems are available at the plant and field levels. Research results have led to practical recommendations for fertilizer or manure applications to maximize crop or pasture production and economic returns. More recent investigations of soil biology and its changes due to different management strategies have revealed some of the mechanisms of nutrient responses by plants, especially in organic or ecological systems. There is extensive information on N fixation that helps explain some rotation effects, and much contemporary research on mychorrizal activity that has expanded our understanding of P dynamics in the plant–soil system. More research is needed on the interactions among nutrients, soil microbes and pathogens, and soil conditions, especially nutrients' dependence on soil moisture and on other soil factors at various levels. Concentration of P deposits in relatively few places in the world, their exploitation for agricultural uses, and their finite nature all point to efficient P use as one of the priority issues we must face in the future.

Finite supplies of fresh water for agriculture and other human uses, plus the strong relationship between soil moisture and the water cycle and water's critical influence on soil nutrient flows and cycles, combine to suggest that future research on water will be one of the most important factors in our successful management of soil nutrition. So important is water that future books on agroecosystems analysis may well focus on water use efficiency as an overriding factor in most areas of the world.

The study of soils and nutrients has involved the water cycle at watershed, regional, and global scales. Agriculture has been identified as a major contributor to nonpoint source contamination of lakes and waterways. Excessive application or poorly managed nutrients in the cropping field or pasture and their escape from confined animal operations represent wasted and expensive resources in agriculture as well as costly impacts to society. To recognize these lost nutrients and animal wastes as valuable resources can lead to more careful, efficient, and economic management of nutrient cycles in agroecosystems, which will lead to more sustainable use of scarce resources and reduced societal costs off the farm.

More research attention is needed on systems at the higher levels of spatial scale beyond the plant and the field. We also need to focus on longer-term issues than single-year costs and returns to fertilizer use on the farm, since soil fertility is important to long-term crop and animal productivity. Recognition by society of the critical nature of soil fertility and productivity to long-term sustainability of human food systems can lead to national legislation and international accords that will reward efficient nutrient use; promote C, N, and other element cycling at the most local level possible; and give financial incentives to farmers who promote and use such practices. Financial incentives for C sequestration are one current example of such legislation. Global initiatives such as the Kyoto accord (which the USA has not signed as of 2003) recognize the importance of global cycles and the urgent need to modify human resource use for the long-term good of all societies (UNFCCC, 2002). Study of nutrient cycles and flows across a spatial scale of interest can enhance our understanding of the connectedness of all systems on the local, regional, and global levels, and can contribute to a credible approach to agroecosystems analysis.

STUDY QUESTIONS

1. Discuss the relative importance of the local and global water, C, N, and P cycles for long-term sustainability of food production.

2. Why has soil biology received relatively less attention than crop yield responses to fertilizer or manure applications? How is this situation changing?

3. Define soil quality and soil health, and describe their importance to crop and livestock production and long-term sustainability of farms and ranches.

4. In what ways can the study of natural ecosystems lead to useful understanding of agroecosystem structure and function? How can this information be useful in the design of future systems?

5. How does the study of nutrient cycles and flows across different levels in a spatial hierarchy aid our understanding of agricultural systems? How does this influence our research priorities?

REFERENCES

Bleken, M.A., and L. A. Bakken. 1997. The nitrogen cost of food production: Norwegian society. Ambio 26:134–142.

Brouwer, F.M., F.E. Godeschalk, P.J.G.J. Hellegers, and H.J. Kelholt. 1995. Mineral balances at farm level in the European Union. Agric. Econ. Re. Inst. (LEI-DLO), Den Haag, the Netherlands.

Caldwell, R.M. 2004. Geographic information systems and landscape analysis. p. 127–146. *In* Agroecosystems analysis. Agron. Monogr. 43. ASA, CSSA, SSSA, Madison, WI.

Cassman, K.G., A. Doberman, and D.T. Walters. 2002. Agroecosystems, nitrogen-use efficiency, and nitrogen management. Ambio 31:132–140.

Correll, D.L. 1983. N and P in soils and runoff from three coastal plain land uses. p. 207–224. *In* R.R. Lowrance et al. (ed.) Nutrient cycling in agricultural ecosystems. Spec. Publ. 23. College of Agric., Univ. of Georgia, Athens.

Crookston, R.K., J.E. Kurle, P.J. Copland, J.H. Ford, and W.E. Lueschen. 1991. Rotational cropping sequence affects yield of corn and soybean. Agron. J. 83:108–113.

Diamond, J. 1997. Guns, germs, and steel: The fates of human societies. W.W. Norton & Co., New York.

Doran, J.W., and A.J. Jones (ed.) 1996. Methods for assessing soil quality. SSSA Spec. Publ. 49. SSSA, Madison, WI.

Doran, J.W., and J.T. Sims. 2002. Sustaining earth and its people. Geotimes, July, p. 5.

Francis, C., L. Salomonsson, G. Lieblein, and J. Helenius. 2004. Serving multiple needs with rural landscapes and agricultural systems. p. 31–48. *In* Agroecosystems analysis. Agron. Monogr. 43. ASA, CSSA, SSSA, Madison, WI.

Galloway, J., E. Cowling, and E. Kessler (ed.) 2002. Reactive nitrogen. Ambio 31:59–196.

Galloway, J.N., E.B. Cowling, J.W. Erisman, J. Wisniewski, and C. Jordan (ed.) 2001. Optimizing nitrogen management in food and energy production and environmental protection. *In* Proc. 2nd Intl. Nitrogen Conf, Potomac, Maryland. A.A. Balkema Publ., Lisse, the Netherlands.

Granstedt, A., and L. Westberg. 1993. Flows of nutrients from agriculture. Aktuellt från landbruksuniversitetet 416. Mark Växter, Swedish Agricultural University, Uppsala.

Hansen, J.F. 1989. Nitrogen balance in agriculture in Denmark and ways of reducing the loss of nitrogen. p. 1–15. *In* Management systems to reduce impacts of nitrates. Elsevier Science Publ., New York.

Hillel, D. 1992. Out of the earth: Civilization and the life of the soil. Univ. California Press, Berkeley.

Isermann, K. 1991. Nitrogen and phosphorus balances in agriculture: A comparison of several western European countries. Intl. Conf. on Nitrogen, Phosphorus, and Organic Matter, Helsingør, Denmark. 13–15 May 1991.

Karlen, D.L., G.E. Varvel, D.G. Bullock, and R.M. Cruse. 1994. Crop rotations for the 21st Century. Adv. Agron. 53:1–43.

Kennedy, A., and J. Doran. 2002. Sustainable agriculture: Role of microorganisms. p. 3116–3126. *In* G. Bitton (ed) Environmental biology. Vol. 6. John Wiley & Sons, New York.

Lampkin, N. 1990. Organic farming. Farming Press, Ipswich, NY.

Lotter, D. 2003. Organic agriculture. J. Sustain. Agric. 21:59–128.

Mäder, P., A. Fliessbach, D. Dubois, L. Gunst, P. Fried, and U. Niggli. 2002. Soil fertility and biodiversity in organic farming. Science 296:1694–1697.

Magdoff, F., L. Lanyon, and B. Liebhardt. 1997. Nutrient cycling, transformations, and flows: Implications for a more sustainable agriculture. Adv. Agron. 60:1–73.

Magdoff, F., and H. van Es. 2000. Building soils for better crops. 2nd ed. Sustainable Handb. Ser. Book 4. Sustainable Agric. Network (SAN), CSREES, USDA, Washington DC.

Olsthoorn, C.S.M. 1992. Minerals in agriculture, 1990. Kvartaalbericht Milieustatistieken, CBS 1(9):16–20.

Rickerl, D., and C. Francis. 2004. Multidimensional thinking: A prerequisite to agroecology. p. 1–18. *In* Agroecosystems analysis. Agron. Monogr. 43. ASA, CSSA, SSSA, Madison, WI.

Sapek, A., and B. Sapek. 1993. Assumed nonpoint water pollution based on the nitrogen budget in Polish agriculture. Water Sci. Technol. 28:483–488.

Sibbesen, E. 1990. Nitrogen, phsphorus and potassium in feed stuff, animal production and manure of Danish agriculture. Tidskr. for Planteavls. Specialserie S 2054. Royal Veterinary and Agricultural University (KVL), Frederiksberg, Denmark.

Soule, J.D., and J.K. Piper. 1992. Farming in nature's image: An ecological approach to agriculture. Island Press, Washington, DC.

UNFCCC. 2002. A guide to the climate convention and its Kyoto protocol [Online]. Available at http://unfccc.int/resource/guideconvkp-p.pdf (verified 1 Oct. 2003). U.N. Climate Change Secretariat, Bonn, Germany.

Yaalov, D.H. 2000. Down to earth: Why soil—and soil science—matters. Nature 407:301.

4 Designing Species-Rich, Pest-Suppressive Agroecosystems through Habitat Management

CLARA I. NICHOLLS AND MIGUEL A. ALTIERI

Division of Insect Biology
University of California
Berkeley, California

Ninety-one percent of the 1.5 billion hectares of cropland worldwide are under annual crops, mostly monocultures of wheat (*Triticum* spp.), rice (*Oryza sativa* L.), maize (*Zea mays* L.), cotton (*Gossypium hirsutum* L.), and soybean [*Glycine max* (L.) Merr.] (Altieri, 1999). This process represents an extreme form of simplification of nature's biodiversity. Monocultures, in addition to being genetically uniform and species-poor systems, advance at the expense of natural vegetation, a key landscape component that provides important ecological services to agriculture such as natural mechanisms of crop protection (Altieri, 1999). Since the onset of agricultural modernization, farmers and researchers have been faced with a major ecological dilemma arising from the homogenization of agricultural systems: an increased vulnerability of crops to insect pests and diseases, which can be devastating when infesting uniform-crop, large- scale monocultures (Adams et al., 1971; Altieri and Letourneau, 1982, 1984). Monocultures may have temporary economic advantages for farmers, but in the long term they do not represent an ecological optimum. Rather, the drastic narrowing of cultivated plant diversity has put the world's food production in greater peril (National Academy of Sciences, 1972; Robinson, 1996).

In this chapter, we explore practical steps to break the nondiverse nature of monocultures and thus reduce their ecological vulnerability by restoring agricultural biodiversity at the field and landscape level. The most obvious advantage of diversification is a reduced risk of total crop failure due to invasions by unwanted species and subsequent pest infestations (Altieri, 1994). The chapter focuses on ways in which biodiversity can contribute to the design of pest-stable agroecosystems by creating an appropriate ecological infrastructure within and around cropping systems. Selected studies reporting the effects of intercropping, cover cropping, weed management, agroforestry, and manipulation of crop-field border vegetation are discussed, with special attention given to understanding the mechanisms underlying pest reduction in diversified agroecosystems. This reflection is fundamental if habitat management through vegetation diversification is to be used effectively as the basis of ecologically based pest management (EBPM) tactics in sustainable agriculture.

BIODIVERSITY IN AGROECOSYSTEMS: TYPES AND ROLES

Biodiversity refers to all species of plants, animals, and microorganisms existing and interacting within an ecosystem, and which play important ecological functions such as pollination, organic matter decomposition, predation or parasitism of undesirable organisms, and detoxification of noxious chemicals (Gliessman, 1998). These renewal processes and ecosystem services are largely biological; therefore, their persistence depends on maintenance of ecological diversity and integrity. When these natural services are lost due to biological simplification, the economic and environmental costs can be quite significant. Economically, in agriculture the burdens include the need to supply crops with costly external inputs, since agroecosystems deprived of basic regulating functional components lack the capacity to sponsor their own soil fertility and pest regulation. Often the costs also involve a reduction in the quality of life of rural communities because of decreased soil, water, and food quality when pesticide, nitrate, or other contamination linked to industrial agriculture occurs (Conway and Pretty, 1991).

Biodiversity in agroecosystems can be as varied as the many crops, weeds, arthropods, or microorganisms involved, according to geographical location, climatic, edaphic, human, and socioeconomic factors. In general, the degree of biodiversity in agroecosystems depends on several features of the agroecosystem. Higher levels of biodiversity are expected in systems that (Altieri, 1994):

- Maintain diversity of vegetation within and around the agroecosystem
- Exhibit temporal and spatial permanence of the various crops within the agroecosystem
- Are subject to low management intensity
- Are not isolated from natural vegetation

The biodiversity components of agroecosystems can be classified according to the role they play in the functioning of cropping systems. Thus, agricultural biodiversity can be grouped as follows (Altieri, 1994; Gliessman, 1998):

- Productive biota—crops, trees, and animals that are chosen by farmers and play a determining role in the diversity and complexity of the agroecosystem
- Resource biota—organisms that contribute to productivity through pollination, biological control, decomposition
- Destructive biota—weeds, insects pests, and microbial pathogens, which farmers aim at reducing through cultural management

The above categories of biodiversity can further be understood as two distinct components (Vandermeer and Perfecto, 1995). The first component, *planned biodiversity*, includes the crops and livestock purposely included in the agroecosystem by the farmer, which will vary depending on the management inputs and crop spatial and temporal arrangements. The second component, *associated biodiversity*, includes all soil flora and fauna, herbivores, carnivores, and decomposers that colonize the agroecosystem from surrounding environments and that will thrive in the agroecosystem depending on its management and structure. The relationship of both types of biodiversity components is illustrated in Fig. 4–1. Planned biodiversity has a direct function, as illustrated by its connection with the ecosystem function box. Associated biodiversity also has a function, but it is mediated through planned biodiversity. Thus, planned biodiversity also has an indirect function, illustrated by the dotted arrow in the figure, which is realized through its influence on the associated biodiversity. For example, the trees in an agroforestry system create shade, which makes it possible to grow only sun-intolerant crops. So, the direct function of this second species (the trees) is to create shade, yet along with the trees might come wasps that seek out the nectar in the tree's flowers. These wasps may in turn be the natural parasitoids of pests that normally attack understory crops. The wasps are part of the associated biodiversity. The trees create shade (direct function) and attract wasps (indirect function) (Vandermeer and Perfecto, 1995).

The optimal behavior of agroecosystems depends on the level of interactions among the various biotic and abiotic components. By assembling a functional biodiversity it is possible to initiate synergisms that subsidize agroecosystem processes by providing ecological services such as the activation of soil biology, the recycling of nutrients, the enhancement of beneficial arthropods and antagonists, and so on, all important components that determine the sustainability of agroecosystems (Altieri and Nicholls, 2000).

The key is to identify the type of biodiversity that is desirable to maintain and/or enhance in order to carry out ecological services, and then to determine the best practices that will encourage the desired biodiversity components. There are

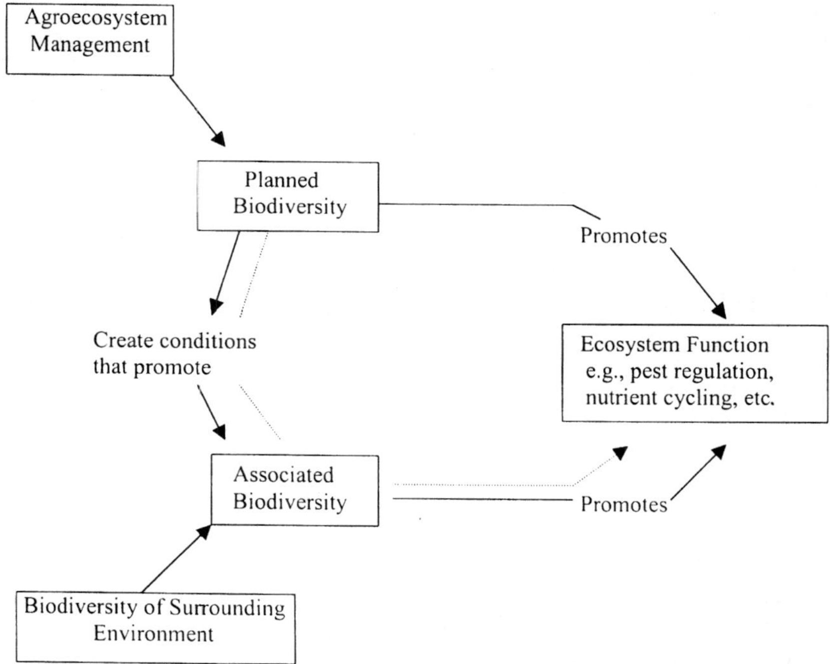

Fig. 4–1. The relationship between planned biodiversity (that which the farmer determines based on management of the agroecosystem) and associated biodiversity and how the two promote ecosystem function. (Modified from Vandermeer and Perfecto, 1995.)

many agricultural practices and designs that have the potential to enhance functional biodiversity, and others that affect it negatively (Fig. 4–2). The idea is to apply the best management practices in order to enhance or regenerate the kind of biodiversity that can best subsidize the sustainability of agroecosystems by providing ecological services such as biological pest control, nutrient cycling, and water and soil conservation. The role of agroecologists should be to encourage those agricultural practices that increase the abundance and diversity of above- and belowground organisms, which in turn provide key ecological services to agroecosystems.

Thus, a key strategy of EBPM should be to exploit the complementarity and synergy that result from the various combinations of crops, trees, and animals in agroecosystems that feature spatial and temporal arrangements such as polycultures, agroforestry systems, and crop–livestock mixtures. In real situations, the exploitation of these interactions involves agroecosystem design and requires an understanding of the numerous relationships among soils, microorganisms, plants, insect herbivores, and natural enemies to guide proper management.

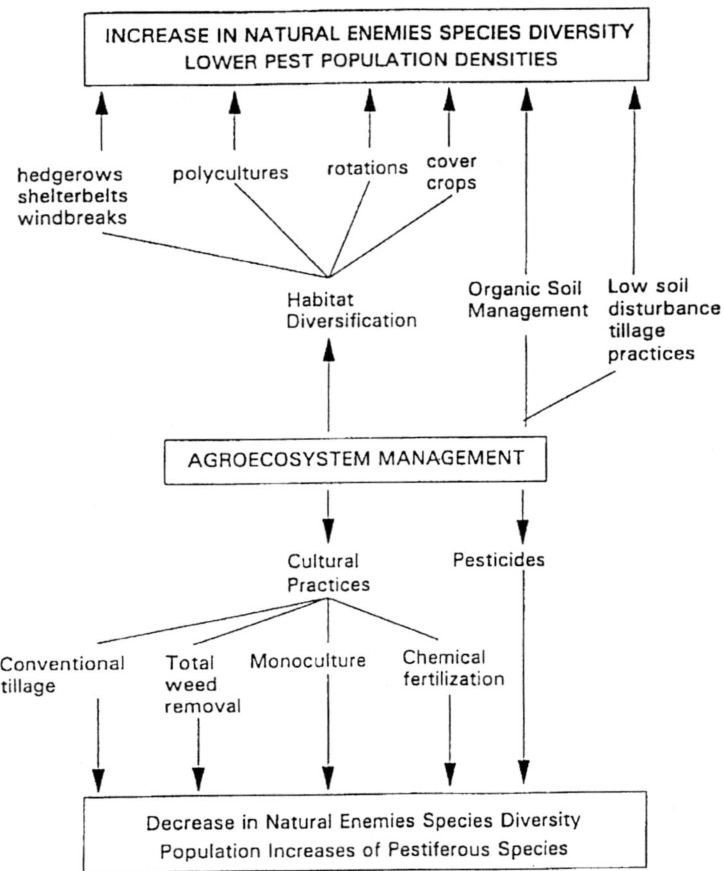

Fig. 4–2. The effects of agroecosystem management and associated cultural practices on the biodiversity of natural enemies and the abundance of insect pests.

LINKING BIODIVERSITY AND AGROECOSYSTEM STABILITY

In general, natural ecosystems appear to be more stable and less subject to fluctuations in populations of the organisms making up the community than are cultivated systems. Ecosystems with higher diversity are more stable because they exhibit:

- Higher resistance, or the ability to avoid or withstand disturbance
- Higher resilience, or the ability to recover following disturbance

The community of organisms becomes more complex when a larger number of different kinds of organisms are included, when there are more interactions among organisms, and when the strength of these interactions increases. As diversity increases, so do opportunities for coexistence and beneficial interference between species that can enhance agroecosystem sustainability (van Emden and Williams, 1974). A diverse system encourages a complex food web, which entails more potential connections and interactions among members, and many alternative paths of energy and material flow through it. For this and other reasons a more complex community exhibits more stable production and less fluctuations in the numbers of undesirable organisms (Power, 1999).

Recent studies conducted in grassland systems suggest that there are no simple links between species diversity and ecosystemic stability. What is apparent is that functional characteristics of component species are as important as the total number of species. The experiments on grassland plots suggest that functionally different roles represented by plants are at least as important as the total number of species in determining processes and services in ecosystems (Tilman et al., 1997).

This latest finding has practical implications for agroecosystem management. If it is easier to mimic specific ecosystem processes rather than to duplicate all the complexity of nature, then the focus should be placed on incorporating a specific biodiversity component that plays a specific role, such as a plant that fixes N, provides cover for soil protection, or harbors resources for natural enemies. In the case of farmers without major economic and resource limitations and who can afford a certain risk of crop failure, a crop rotation or a simple crop association may be all it takes to achieve a desired level of stability. But in the case of resource- poor farmers, where crop failure is intolerable, highly diverse polyculture systems would probably be the best choice. The obvious reason is that the benefit of complex agroecosystems is low risk; if a species falls to disease, pest attack, or weather, another species is available to fill the void and maintain full use of resources. Thus, there are potential ecological benefits to having several species in an agroecosystem: compensatory growth, full use of resources and nutrients, and pest protection (Ewel, 1999).

Plant Diversity and Insect Pest Regulation

Throughout the years, many ecologists have conducted experiments testing the hypothesis that decreased plant diversity in agroecosystems, which allows greater chance for invasive species to colonize, subsequently leads to enhanced herbivorous insect pest abundance. Many of these experiments have shown that mixing certain species with the primary host of a specialized herbivore gives a fairly consistent result: specialized species usually exhibit higher abundance in monoculture than in diversified crop systems (Andow, 1991).

Several reviews have been published documenting the effects of within-habitat diversity on insects (Altieri and Letourneau, 1984; Risch et al., 1983). Two main ecological hypotheses, (i) the natural enemy hypothesis and (ii) the resource concentration hypothesis, have been offered to explain why insect communities in agroecosystems can be stabilized by constructing vegetational architectures that support natural enemies and/or directly inhibit pest attack. The literature is full of examples of experiments documenting that diversification of cropping systems often leads to reduced herbivore populations. In the review by Risch et al. (1983), 150 published studies of the effect of diversifying an agroecosystem on insect pest abundance were summarized; 198 total herbivore species were examined in these studies. Fifty-three percent of these species were found to be less abundant in the more diversified system, 18% were more abundant in the diversified system, 9% showed no difference, and 20% showed a variable response. In another analysis of 50 studies, it was concluded that monophagous (specialist) insects are more susceptible to crop diversity than polyphagous insects. The author cautioned about the increased risk of pest attack if the dominant herbivore fauna in a given agroecosystem is polyphagous (Andow, 1991). The reduction in pest numbers for monophagous insects was almost twice (53.5% of the case studies showed lowered numbers in polycultures) that for polyphagous insects (33.3% of the cases).

Both empirical data and theoretical arguments suggest that differences in pest abundance between diverse and simple annual cropping systems can be explained by the movement and reproductive behavioral responses of herbivores when confronted with plant diversity and, in many cases, by mortality imposed by natural enemies. Of 35 insect pest species investigated in one study, the majority in the orders Lepidoptera, Coleoptera, and Homoptera, natural enemy action accounted for 30% of the control mechanisms of the various pests, and the remaining species were controlled by a variety of factors, including lowered resource concentration, trap-cropping, diversionary mechanisms, and plant physical obstruction (Barbosa, 1998).

Recent reviews concerned with agroecology, habitat management, and conservation biological control overwhelmingly state that higher pest losses should be expected in more vulnerable ecosystems, usually mechanized, large-scale monocultures (Altieri, 1994; Barbosa, 1998; Pickett and Bugg, 1998). Such systems represent highly disturbed systems exhibiting ecological conditions that may be more susceptible to colonization by invasive species. Herbivores with a narrower host range are more likely to colonize crops grown in pure stands and thus attain pest states in simplified agroecosystems (Smith and McSorely, 2000). Moreover, as a result of frequent and intense disturbance regimes, monocultures are difficult environments for natural enemies to colonize and survive in; thus, predators and parasitoids reach low abundance levels and exhibit poor effectiveness in such systems. The ubiquity of pesticide use negatively impacts natural enemies, and high rates of synthetic chemical fertilizer may render crops more susceptible to pests. Effects of transgenic crops on nontarget organisms will not be as localized or as transient as initially anticipated. Rather, studies suggest that the effects of transgenic crops might spread via wind or trophic webs, and might persist in the soil, in many cases compounding pest problems (Marvier, 2001).

DESIGNING BIODIVERSE PEST-SUPPRESSIVE AGROECOSYSTEMS

Monoculture agriculture is a futile attempt to impose agronomic simplicity on ecosystems that are inherently complex and possess high biotic intricacy. Accordingly, many agroecologists have proposed that a better land use management strategy is to imitate the structure and function of the natural communities of each region (Ewel, 1986). Successional communities offer several traits of potential value to agriculture (Soule and Piper 1992; Ewel 1999):

- High resistance to pests and diseases
- High nutrient retention and recycling capacity
- High levels of biodiversity and positive synergisms among biotic components
- Higher presence of perennials and level of ecosystem permanence

A strategy to bring such benefits to agricultural systems is to use successional ecosystems as templates for the design of agroecosystems, a strategy that has been used for centuries by traditional tropical small farmers in the design of polycultures, agroforestry, and complex home gardens. In modern agricultural systems, the same strategy can be used, with the following key ecological guidelines:

- Increase species in time and space through multiple cropping and agroforestry designs.
- Increase genetic diversity through variety mixtures, multilines, and use of local germplasm and varieties exhibiting horizontal resistance.
- Include an improved fallow through legume-based rotations, use of green manures, cover crops, and/or livestock integration.
- Enhance landscape diversity with vegetationally diverse crop field boundaries or by creating a mosaic of agroecosystems and maintaining areas of natural or secondary vegetation.

Recent case studies confirm that adoption of some form of diversification following the key agroecological principles outlined above can lead to enhanced pest regulation. The next sections illustrate how many of these studies have transcended the research phase and have found wide applicability to regulate specific pests.

Rice Polyculture. Researchers working with farmers in ten townships in Yumman, China, covering an area of 5350 hectares, encouraged farmers to switch from rice monocultures to planting variety mixtures of local rice with hybrids. Enhanced genetic diversity reduced blast incidence by 94% and increased total yields by 89%. By the end of 2 yr, it was concluded that fungicides were no longer required (Zhu et al., 2000; Wolfe, 2000).

Push-Pull. In Africa, scientists at ICIPE developed a habitat management system to control stemborers and striga (*Striga* spp.), which uses two kinds of crops that are planted together with maize: a plant that repels borers (the push) and another that attracts (pulls) them acting as a trap crop (Khan et al., 2000). The push–pull system has been tested on more than 450 farms in two districts of Kenya and has now been released for uptake by the national extension systems in East Africa. Participating farmers in the breadbasket of Trans Nzoia are reporting a 15 to 20% increase in maize yield. In the semiarid Suba district—plagued by both stemborers and striga—a substantial increase in milk yield has occurred in the last 4 yr, with farmers now able to support grade cows (*Bos taurus*) on the fodder produced. When farmers plant maize, napier (*Pennisetum purpureum* Schumach.), and desmodium (*Desmodium* spp.) together, a return of $2.30 for every dollar invested is made, as compared with only $1.40 obtained by planting maize as a monocrop. Two of the most useful trap crops that pull in the borers' natural enemies are napier grass (*Pennisetum purpureum*) and Sudan grass (*Sorghum vulgare* var. *sudanense* Hitchc.), both important fodder plants; these are planted in a border around the maize. Two excellent borer-repelling crops which are planted between the rows of maize are molasses grass (*Melinis minutifolia* P. Beauv.), which also repels ticks, and the leguminous silverleaf [*Desmodium uncinatum* (Jacq.) DC.]. This plant can also suppress the parasitic weed striga by a factor of 40 compared with maize monocrops. Its N-fixing ability increases soil fertility; and it is an excellent forage. As an added bonus, sale of desmodium seed is proving to be a new income- generating opportunity for women in the project areas.

Flower Strips and Beetle Banks. Several researchers have introduced flowering plants as strips within crops as a way to enhance the availability of pollen and nectar, necessary for optimal reproduction, fecundity and longevity of many natural enemies of pests. *Phacelia tanacetifolia* Benth. strips have been used in wheat, sugarbeet (*Beta vulgaris* L. subsp. *vulgaris*), and cabbage (*Brassica oleracea* L.), leading to enhanced abundance of aphidophagous predators, especially syrphid flies, and reduced aphid populations. In England, in an attempt to provide suitable overwintering habitat within fields for predators of cereal aphids, researchers created "beetle banks" sown with perennial grasses such as *Dactylis glomerata* L. and *Holcus lanatus* L.). When these banks run parallel with the crop rows, great enhancement of predators (up to 1500 beetles per square meter) can be achieved in only 2 yr (Landis et al., 2000).

Flowering Undergrowth in Perennial Cropping. In perennial cropping systems the presence of flowering undergrowth enhances the biological control of a series of insect pests. The beneficial insectary role of *Phacelia* species in apple (*Malus domestica* Borkh.) orchards was well demonstrated by Russian and Canadian researchers more than 30 yr ago (Altieri, 1994). In California organic vineyards, the incorporation of flowering summer cover crops (buckwheat [*Fagopyrum esculentum* Moench] and sunflower [*Helianthus annuus* L.]) led to enhanced populations of natural enemies, which in turn significantly reduced the numbers of leafhoppers and thrips (Nicholls et al., 2000).

Weeds as Plant Diversity. In Washington State, researchers reported that organic apple orchards managed with lower inputs and that retained some level of plant diversity in the form of weeds mowed as needed gave similar apple yields to conventional and integrated orchards. Their data showed that the organic system ranked first in environmental and economic sustainability as this system exhibited higher profitability, greater energy efficiency, and lower negative environmental impact (Reganold et al., 2001).

Multistrata Shade-Grown Coffee Systems. In Central America, Staver et al. (2001) designed pest-suppressive multistrata shade-grown coffee systems, selecting tree species and associations, density and spatial arrangement, as well as shade management regimes, with the main goal of creating optimum shade conditions for pest suppression. For example, in low- elevation coffee zones, 35 to 65% shade promotes leaf retention in the dry seasons and reduces the pathogen *Cercospora coffeicola* Berk. & Cke., weeds, and the insect citrus mealybug [*Planococcus citri* (Rossi)]. At the same time, the shade enhances the effectiveness of microbial and parasitic organisms without contributing to increased rust (*Hemileia vastatrix* Berke & Br.) levels or reducing yields.

Importance of Field Margins. Several entomologists have concluded that the abundance and diversity of predators and parasites within a field are closely related to the nature of the vegetation in the field margins. There is wide acceptance of the importance of field margins as reservoirs of the natural enemies of crop pests, although, depending on plant composition, certain hedgerows may also harbor pests. Many studies have demonstrated increased abundance of natural enemies and more effective biological control where crops are bordered by wild vegetation from which natural enemies colonize. Parasitism of the armyworm, *Pseudaletia unipunctata* (Hayworth), was significantly higher in maize fields embedded in a complex landscape than in maize fields surrounded by simpler habitats. In a 2-yr study researchers found higher parasitism of larvae of the lepidopteran pest, *Ostrinia nubilalis* (Hübner) by the parasitoid *Eriborus terebrans* (Gravenhorst) in edges of maize fields adjacent to wooded areas, than in field interiors (Landis et al., 2000). Similarly, in Germany, parasitism of rape pollen beetle (*Meligethes aeneus* F.) was about 50% at the edge of the fields, while at the center of the fields parasitism dropped significantly to 20% (Thies and Tscharntke, 1999).

Vegetational Corridors and Arthropod Diversity. One way to introduce the beneficial biodiversity from surrounding landscapes into large-scale monocultures is by establishing vegetationally diverse corridors that allow the movement and distribution of useful arthropod biodiversity into the center of monocultures. Nicholls et al. (2001) established a vegetational corridor that connected to a riparian forest and cut across a vineyard monoculture in northern California. The corridor allowed natural enemies emerging from the riparian forest to disperse over large areas of otherwise monoculture vineyard systems. The corridor provided a constant supply of alternative food for predators, effectively decoupling predators from a strict dependence on grape herbivores and avoiding a delayed colonization of the vineyard. This complex of predators continuously circulated into the vineyard interstices and established a set of trophic interactions leading to a natural enemy enrichment, which led to lower numbers of leafhoppers and thrips on vines located up to 30 to 40 m from the corridor.

All of the above examples constitute forms of habitat diversification that provide resources and environmental conditions suitable for natural enemies. The challenge is to identify the type of biodiversity that is desirable to maintain and/or enhance in order to carry out ecological services of pest control and then to determine the best practices that will encourage such desired biodiversity components. A few guidelines need to be considered when implementing habitat management strategies (Landis et al 2000):

- Select the most appropriate plant species
- Determine the most beneficial spatial and temporal arrangement of such plants, within and/or around the fields.
- Consider the spatial scale at which the habitat enhancement operates (e.g., field or landscape level)
- Understand the predator–parasitoid behavioral mechanisms influenced by the habitat manipulation.
- Anticipate potential conflicts that may emerge when adding new plants to the agroecosystem (i.e., in California, blackberries (*Rubus* spp.) around vineyards increase populations of wasps (*Anagrus epos* Girault), a parasitoid of the grape leafhopper (*Erythroneura* spp.), but can also enhance abundance of the sharpshooter, which serves as a vector of Pierce's disease).
- Develop ways in which the added plants do not upset other agronomic management practices, and select plants that have multiple effects, such as improving pest regulation while at the same time contributing to soil fertility and weed suppression.

CONCLUSIONS

The instability of agroecosystems, which is manifested as the worsening of most insect pest problems, is increasingly linked to the expansion of crop monocultures at the expense of the natural vegetation, thereby decreasing local habitat diversity (Altieri, 1994). Plant communities that are modified to meet the special needs of humans become subject to heavy pest damage. Generally, the more intensely such communities are modified, the more abundant and serious the pests. The inherent self-regulation characteristics of natural communities are lost when humans modify such communities through the shattering of the fragile thread of community interactions. Agroecologists maintain that restoring the shattered elements of the community homeostasis through the addition or enhancement of biodiversity (Gliessman, 1999; Altieri, 1999) can repair this breakdown.

A key strategy in sustainable agriculture is to reincorporate diversity into the agricultural landscape through various cropping designs. Emergent ecological properties develop in diversified agroecosystems that allow the system to function in ways that maintain soil fertility, crop production, and pest regulation. The main approach in ecologically based pest management is to use management methods that increase agroecosystem diversity and complexity as a foundation for establishing beneficial interactions that keep pest populations in check (Altieri and

Nicholls, 2000). This is particularly important in underdeveloped countries where sophisticated inputs are either not available or may not be economically or environmentally advisable, especially in the case of resource-poor farmers.

As argued in this chapter, agroecosystems that mimic the structure and functional complexity of nature confer an important degree of pest protection. However, diverse and complex agroecosystems are hard to manage, and their implementation may run counter to current economic forces that promote farm specialization. Nevertheless, new agroecosystems are urgently needed worldwide in an era of deteriorating environmental quality, biodiversity reduction, heavy reliance on nonrenewable resources, and escalating input costs. This approach to agriculture will only be practical if it is economically sensible and can be carried out within the constraints of a fairly normal agricultural management system. However, given the trend toward large-scale, monoculture production units throughout the world (USDA, 1973), objectively there is not much room left for a fair implementation of a regional insect-habitat management program. Emerging biotechnological approaches, such as transgenic crops deployed in more than 40 million hectares in 2000, are leading agriculture towards further specialization, and the potential effects of transgenic crops on nontarget beneficial organisms is of concern to biological control practitioners (Rissler and Mellon, 1996; Altieri, 2000; Marvier, 2001). Regardless, habitat management may not always demand a radical change in farming, as illustrated by the relative ease with which beetle banks, flowering strips, or corridors can be introduced into cropping systems, and thus bringing biological control benefits to farmers (Landis et al., 2000).

When properly implemented, habitat management leads to establishment of the desired type of plant biodiversity and the ecological infrastructure necessary for attaining optimal natural enemy diversity and abundance. Such diversity may not always warrant total pest regulation; therefore, at times the action of such enemies might have to be complemented with augmentative releases of predators or parasites and/or application of entomopathogens. This may be especially true in the initial stages of the conversion from conventionally managed systems to agroecological management (Vandermeer, 1995; Landis et al., 2000).

Long-term maintenance of diversity requires a management strategy that considers regional biogeography and landscape patterns, as well as design of environmentally sound agroecosystems above purely economic concerns. This is why several authors have repeatedly questioned whether the pest problems of modern agriculture can be ecologically alleviated within the context of the present capital-intensive structure of agriculture. Many problems of modern agriculture are rooted within that structure and thus require the consideration of major social change, land reform, redesign of machinery, research, and extension reorientation in the agricultural sector to increase the possibilities of improved pest control through vegetation management. Whether the potential and spread of ecologically based pest management is realized will depend on policies, attitude changes on the part of researchers and policy makers, existence of markets for organic produce, and also the organization of farmer and consumer movements that demand a more healthy and viable agriculture and food system.

STUDY QUESTIONS

1. Compare the characteristics of a "natural" ecosystem and an agricultural monoculture. What problems and risks arise from the simplification of natural systems into agricultural monocultures?

2. Does species diversity guarantee ecosystem stability? Why or why not?

3. Discuss a successional ecosystem that could serve as an agroecosystem model in your region. How does it incorporate the key agroecological principles outlined in this chapter?

4. Discuss the mechanisms of pest suppression in your agroecosystem model.

REFERENCES

Adams, M.W., A.H. Ellingbae, and E.C. Rossineau. 1971. Biological uniformity and disease epidemics. BioScience 21:1067–1070.

Altieri, M.A. 1994. Biodiversity and pest management in agroecosystems. Haworth Press, New York.

Altieri, M.A. 1999. The ecological role of biodiversity in agroecosystems. Agric. Ecosyst. Env. 74:19–31.

Altieri, M.A. 2000. The ecological impacts of transgenic crops on agroecosystem health. Ecosyst. Health 6:13–23.

Altieri, M.A., and D.K. Letourneau. 1982. Vegetation management and biological control in agroecosystems. Crop Protect. 1:405–430.

Altieri, M.A., and D.K. Letourneau. 1984. Vegetation diversity and outbreaks of insect pests. Crit. Rev. Plant Sci. 2:131–169.

Altieri, M.A., and C.I. Nicholls. 2000. Applying agroecological concepts to development of ecologically based pest management systems. p. 14–19. *In* Proc. Workshop Professional Societies and Ecological Based Pest Management Systems. National Research Council, Washington, DC.

Andow, D.A. 1991. Vegetational diversity and arthropod population response. Annu. Rev. Entomol. 36:561–586.

Barbosa, P. 1998. Conservation biological control. Academic Press, San Diego.

Conway, G.R., and J.N. Pretty. 1991. Unwelcome harvest: Agriculture and pollution. Earthscan, London.

Ewel, J.J. 1986. Designing agricultural ecosystems for the humid tropics. Annu. Rev. Ecol. Syst. 17:245–71.

Ewel, J.J. 1999. Natural systems as models for the design of sustainable systems of land use. Agrofor. Syst. 45:1–21.

Gliessman, S.R. 1999. Agroecology: Ecological processes in agriculture. Ann Arbor Press, MI.

Khan, Z.R., J.A. Pickett, J. van der Berg, and C.M. Woodcock. 2000. Exploiting chemical ecology and species diversity: Stemborer and *Striga* control for maize in Africa. Pest Manage. Sci. 56:1–6.

Landis, D.A., S.D. Wratten, and G.A. Gurr. 2000. Habitat management to conserve natural enemies of arthropod pests in agriculture. Annu. Rev. Entomol. 45:175–201.

Marvier, M. 2001. Ecology of transgenic crops. Am. Sci. 89:160–167.

National Academy of Sciences. 1972. Genetic vulnerability of major crops. NAS, Washington, DC.

Nicholls, C.I., M.P. Parrella, and M.A. Altieri. 2000. Reducing the abundance of leafhoppers and thrips in a northern California organic vineyard through maintenance of full season floral diversity with summer cover crops. Agric. For. Entomol. 2:107–113

Nicholls, C.I., M.P. Parrella, and M.A. Altieri. 2001. The effects of a vegetational corridor on the abundance and dispersal of insect biodiversity within a northern California organic vineyard. Landscape Ecol. 16:133–146.

Pickett, C.H., and R.L. Bugg. 1998. Enhancing biological control: Habitat management to promote natural enemies of agricultural pests. Univ. of California Press, Berkeley.

Power, A.G. 1999. Linking ecological sustainability and world food needs. Environ. Dev. Sustainability 1:185–196.

Reganold, J.P., J.D. Glover, P.K. Andrews, and H.R. Hinman. 2001. Sustainability of three apple production systems. Nature 410:926–930.

Risch, S.J., D. Andow, and M.A. Altieri. 1983. Agroecosystem diversity and pest control: Data, tentative conclusions, and new research directions. Environ. Entomol. 12:625–629.

Rissler, J., and M. Mellon. 1996. The ecological risks of engineered crops. MIT Press, Cambridge, MA.

Robinson, R.A. 1996. Return to resistance: Breeding crops to reduce pesticide resistance. Ag. Access, Davis, CA.

Smith, H.A., and R. McSorley. 2000. Intercropping and pest management: A review of major concepts. Am. Entomol. 46:154–161.

Soule, J.D., and J.K. Piper. 1992. Farming in nature's image. Island Press, Washington, DC.

Staver, C., F. Guharay, D. Monterroso, and R.G. Muschler. 2001. Designing pest-suppressive multistrata perennial crop systems: Shade grown coffee in Central America. Agrofor. Syst. 53:151–170.

Thies, C., and T. Tscharntke. 1999. Landscape structure and biological control in agroecosystems. Science 285:893–895.

Tillman, D., D. Wedin, and J. Knops. 1996. Productivity and sustainability influenced by biodiversity in grassland ecosystems. Nature 379:718–720.

USDA. 1973. Monocultures in agriculture, causes and problems. Report of the Task Force on Spatial Heterogeneity in agricultural landscapes and enterprises. U.S. Gov. Print. Office, Washington, DC.

Van Driesche, R.G., and T.S. Bellows, Jr. 1996. Biological control. Chapman and Hall, New York.

van Emden, H.F., and G.F. Williams. 1974. Insect stability and diversity in agroecosystems. Annu. Rev. Entomol. 19:455–75.

Vandermeer, J. 1995. The ecological basis of alternative agriculture. Annu. Rev. Ecol. Syst. 26:210–224.

Vandermeer, J., and I. Perfecto. 1995. Breakfast of biodiversity: The truth about rain forest destruction. Food First Books, Institute for Food and Development Policy, Oakland, CA.

Wolfe, M. 2000. Crop strength through diversity. Nature 406:681–682.

Zhu, Y., H. Fen, Y. Wang, Y. Li, J. Chen, L. Hu, and C.C. Mundt. 2000. Genetic diversity and disease control in rice. Nature 406:718–772.

5 Whole-Farm Planning and Analysis

RHONDA R. JANKE

Department of Horticulture, Forestry, and Recreation Resources
Kansas State University
Manhattan, Kansas

Farms are economic units, even though farms as agroecosystem units may not make sense because they often cross agroecosystem boundaries (Fig. 5–1). However, one needs to look at the farm level carefully to understand aggregate processes occurring at hierarchical levels above the farm. We can at least partially understand the farm by looking at the component pieces—the fields, windbreaks, and cattle lots. If one uses a strict definition or concept of agroecosystem as a physical unit with a definite boundary within which one can look at interactions among components and flows into and out of the ecosystem, then farms do make sense.

As landscape units, today's farms are not always contiguous pieces of land. Many management units, such as crop fields, livestock operations, and noncropped areas have few connecting points or physical boundaries and minimal interaction. Many larger farms are made up of a unit of land that may be "the home place," along with rented acres in units that may be in several places within one county or even in several counties. Similarly, if we try to define a farm as a unit within a particular landscape, it leads to frustration because some farms have pieces of land located within several watersheds.

As social structures, farms make sense as units of study and observation. In a social hierarchy that includes national, regional, state, and local forms of government and decision making, a farm is the unit at which specific decisions are made about land and other resource use. As planning units, farms also make sense. The planning is often a complex process, and decisions are made by multiple people, usually within the same family, but often crossing generational boundaries.

This chapter will examine several tools and approaches to whole-farm planning and analysis, some of which fall within the definition of ecosystem analysis. In a literature review of indicators of sustainability for whole-farm planning (Freyenberger et al., 1997), 83 papers were classified according to seven different categories, including: (i) orientation—farmer or academic; (ii) approach—basic, applied, theoretical, conceptual, or management; (iii) orientation—environment, economic, social/community, whole farm, or global; (iv) scale—micro or macro; (v) degree of whole-farm planning emphasis; (vi) applicability—farm/community, state/national, international; and (viii) source of article—author's institutional or research base. That literature review will not be repeated here, but this area of study will be re-

---

Hierarchical Diagram

Global

Continent

Bioregion

River Basin/Subregion

Watershed/Community/Landscape

Farm/Management Unit

Field/Farm Enterprises

Subfield Units

Fig. 5–1. How does the farm as a management unit fit within the hierarchy of agroecosystems?

viewed from several perspectives, including both qualitative and quantitative approaches. Planning tools will be summarized, along with ecological assessment, analysis, and monitoring activities. Some of these tools are especially useful for farmers, and some were designed primarily for use by researchers and policy planners. We will see in the summary that diverse tools are necessary, and a combination of approaches is needed to provide information to both farmers and nonfarmers in order to achieve a comprehensive understanding of farms within an agroecosystem framework.

SOIL AND WATER AT DIFFERENT SCALES

Soil and water are probably the two most important natural resources on a farm. It is not surprising that among the available whole-farm planning tools (Janke and Freyenberger, 1997), the most common types are those used for soil C and soil nutrient management. Tools range from very specific models focused on only one nutrient, such as P, to those that calculate macronutrient budgets for the whole farm. In contrast to this a recent compilation of research papers entitled *Soil and Water Quality at Different Scales* (Finke et al., 1998) included only one paper out of 35 in the monograph that mentioned farm-scale measurement of nutrients (Hack-ten Broeke and de Groot, 1998). This study was conducted using specialized equipment on an experimental farm and didn't consider the complexities of a "real" farm. Many of the experiments and models in this body of research looked at the landscape and then at the field-level or land use characteristics. This raises the question once again: does the farm as an agroecosystem make sense?

It appears that farms do not make sense as part of a strictly hierarchical physical structure, particularly if we are only studying one element, nutrient, or water. However, farms as management units and as decision making units that determine how fields within a landscape will be used cannot be ignored. We also cannot ignore the fact that farms are complex entities, where owners and managers must constantly juggle natural resource and ecosystem goals with economic realities and limitations.

ECONOMIC ANALYSIS

Whole-farm planning as a term was originally used by agricultural economists. Farms as economic units can be summarized in business terms, such as net worth, debt/asset ratio, and cash flow. Units within the farm can be separated out into enterprise budgets. In the review of whole-farm planning tools (Janke and Freyenberger, 1997), the only tools that seemed to be more common than nutrient budgeting tools were economic planning tools. The range of software, spreadsheets, and technical assistance available is vast, and ranges in price go from quite expensive to practically free. However, the number of farmers who take advantage of these resources for financial planning is surprisingly small. Many will use specialized assistance, such as market forecasts for futures trading, but few use whole-farm planning packages such as KSU Farm Analyst Program.

The relationship between economic analysis and agroecosystem analysis is twofold. The first is a minor point and has to do with a turf battle between agricultural economists and natural resource scientists. In the past decade, the term whole-farm planning has been used widely by the natural resource community, largely in the context of sustainable agriculture projects. In some universities, the agricultural economists resent what they see as a cooption of the term. In the future, I hope that the term can be used freely by both groups, with the subdefinitions of the term clear to their respective audiences.

The second point is that these two groups need to work together for successful whole-farm planning and for better resource management from an agroecosystems perspective. For example, a farmer discovers a serious water quality problem on his or her farm, and traces the source back to a specific cattle feeding area. This feedlot has drainage into a channel that flows directly to the stream, with no riparian filter. The farmer then needs to do an economic analysis and to explore both the economic cost and the water quality benefits of several options. One option is to install a riparian buffer or filter. Another, more expensive option might be to move the feeding area to a new location. A third option, which could be appropriate for a farmer nearing retirement, is to stop feeding the cattle and plant the lot to grass.

FARMING SYSTEMS RESEARCH AND EXTENSION (FSR/E)

Early whole-farm planning models can be found in the international agricultural development literature. About 30 yr ago, agricultural economists, agronomists, and others studying farms in an international context developed several tools in what is now called Farming Systems Research/Extension (FSR/E) (Robotham and McArthur, 2001). Early FSR/E approaches included farming systems diagrams (Fig. 5–2), with illustrations of the household, crops, livestock, and their subcomponents, to help researchers from other climates and cultures better understand how best to focus research efforts. Goals were to help farmers increase yield, profit, or crop–livestock integration for increased efficiency. Unlike the economic planning tools that provide useful information directly to farmers, the FSR/E diagrams largely served to educate researchers, with the goal to better focus resources to achieve research objectives. This resulted in improved effectiveness of research and extension efforts, compared with previous Green Revolution efforts that sometimes were carried out in a "one size fits all" manner.

Later FSR/E approaches were broader, and techniques such as rapid rural appraisal and others were used to create diagrams that included not only the farm and farm subunits, but also the interaction of the farm with the larger landscape, and within the community context. This systems approach has largely been adopted by sustainable agriculture researchers, and it has improved the quality of the research and the usefulness to the end users. However, the diagrams, though derived in a participatory fashion with farmer input and feedback, are still simplistic in nature and benefit researchers more than farmers.

In the 1980s, this approach was adapted, and an ecosystems approach to planning that was useful specifically to grassland farmers and ranchers was designed and taught directly to this audience (see e.g., Binham and Savory, 1990). Courses include not only a conceptual diagram that begins with goal setting, but also a thorough explanation of ecosystem processes on the farm, such as C capture (through photosynthesis) and cycling, nutrient cycling, and water cycling. Specific worksheets allow farmers to apply these concepts, including pasture stocking rates, crop and pasture rotation diagrams, and an economic analysis of the whole farm and subunits within the farm using a technique called gross margin analysis (Levins, 1996).

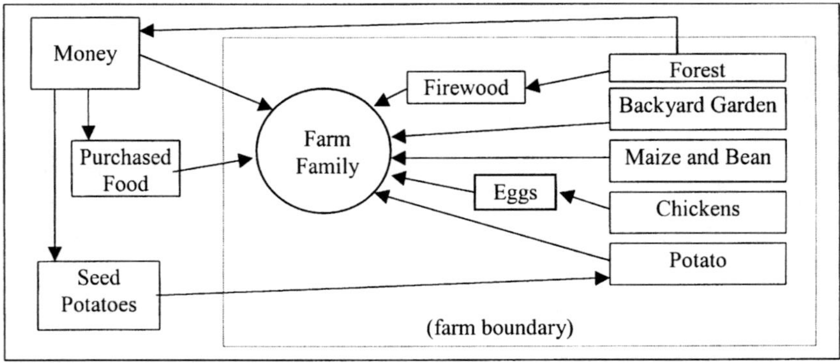

Fig. 5–2. An example of a farm diagram typical of Farming Systems Research/Extension (FSR/E): a subsistence farm.

SUSTAINABILITY INDICES

At about this same time, more complex analyses were developed to evaluate the sustainability of farming and production systems (e.g., Bellows, 1994; Muller, 1994; Stokle et al., 1994; Lightfoot and Noble, 2001). These frameworks included more complexity, such as air, water, and soil quality indicators; energy use efficiency; quality of life; diversity; recycling; and productivity and profitability. They also contained more specific indicators than their predecessors in the FSR/E literature but were still largely conceptual. In some cases, specific farming systems have been used to illustrate how these indicators might be quantified (Fig. 5–3; Lightfoot and Noble, 2001), but generalized case studies of typical farming systems within a region are used rather than individual farms. These frameworks also are most useful to outside observers and might be used by policy makers to determine if programs should be created to provide incentives for farmers to change cropping or farming practices.

A more detailed system for evaluating sustainability on an individual farm has been illustrated by Taylor et al. (1993). In their index, 33 specific questions were answered or scored in areas that include insect, disease, and weed control practices; soil fertility management; and erosion control. The scores were adjusted or weighted, and then the total score was compared with the best possible total score. Similar systems are in use now and are used as a marketing tool by grocery stores (IPM program, Wegmans Food Markets, Rochester, NY), groups of growers (The Food Alliance, Stemitt), and private certifiers interested in promoting ecolabeled products

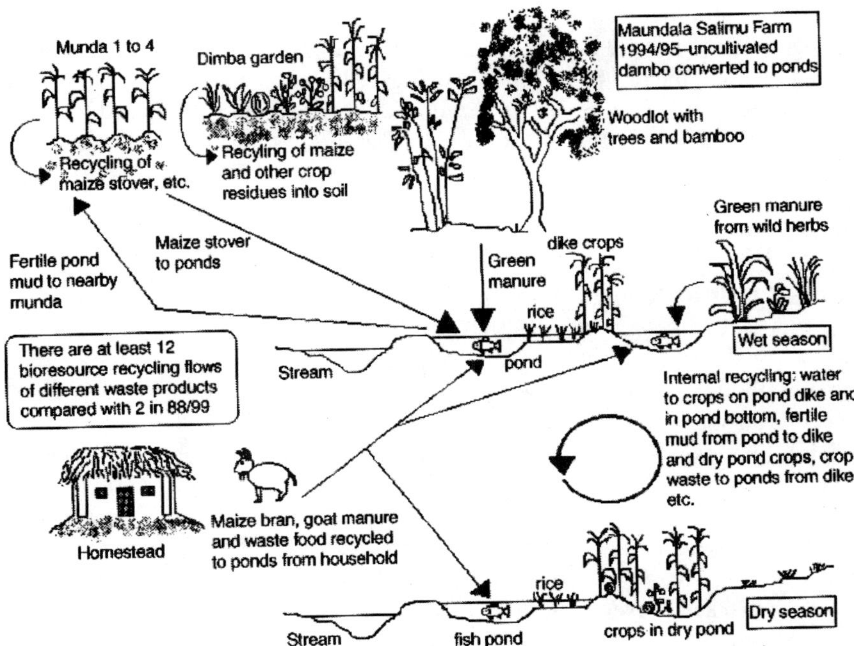

Fig. 5–3. Bio-Resource Flow Model of farming system after transformation, Zomba District, Malawi (from Lightfoot and Noble, 2001, p. 17).

(e.g., "Better Banana" program, Rainforest Alliance, New York). These and other ecolabeling programs are reviewed and critiqued by the Consumers Union (see www.eco-labels.org).

Some of these programs are better than others, as rated by Consumers Union for characteristics such as transparency, consistency, and degree of qualitative vs. quantitative farm evaluation methods. These labels and certifications constitute an interesting example of a whole-farm analysis system that provides not only data, but potential increased economic returns for the farmers as a result of their efforts.

Some components of these frameworks have been promoted to farmers for their use, to aid in decision making on the farm. For example, in the 1980s and 1990s, there was a resurgence of literature and research projects aimed at quantifying such attributes as soil quality (Arshad and Coen, 1992; Granatstein and Bezdicek, 1992; Hunt and Gilkes, 1992; Karlen et al., 1992; Parr et al., 1992). In some cases, specific indicators such as the presence of soil microorganisms (Visser and Parkinson, 1992) and soil invertebrates (Stork and Eggleton, 1992) have been suggested. Water quality monitoring at the farm and community level has also been proposed and explored (Deutsch et al., 1998; IOWATER, 2001; Janke, 2003), but users of these monitoring programs have not attempted to put their data into a sustainability index framework.

A unique approach proposed in Minnesota, The Monitoring Toolbox, encourages farmers to monitor not only soil and water quality and financial indicators, but also birds, frogs and toads, and quality of life, using both qualitative and quantitative methods (Ness, 1998). This doesn't result in a sustainability index per se, but it does allow farmers to see, experience, and quantify some factors on their farms that might be called indicators of sustainability. Frog species diversity is not a direct measurement of the sustainability of a farming practice, or even a direct measurement of the result of a specific practice, such as an increase or decrease in soil or water quality. However, some wildlife species, such as frogs and birds are indicators of the condition of the overall habitat for wildlife. Because species like frogs spend part or all of their life in water, their health and species diversity might also serve as an indicator of water quality on a farm and how it applies to human health.

CURRENT APPROACHES TO NATURAL RESOURCE ASSESSMENT ON FARMS

In addition to the frameworks and approaches reviewed above, there are tools that do not generate an index but are designed to provide the farmer with valuable information or feedback about his or her farm, so that corrective action can be taken if needed.

Some of these include the nutrient management tools mentioned briefly above that allow a farmer to calculate nutrient budgets and determine if there are deficiencies or excesses (see e.g., the Nutrient Management Yardstick used in Minnesota; Janke and Freyenberger, 1997). In the 1980s, a tool was developed called Farm-A-Syst, that is still evolving to assess risk on farms, both environmental and personal health risks. A scorecard is used to primarily look at areas near the farm home, such as well construction and well water quality, pesticide and fertilizer stor-

Table 5–1. Scorecard results from 21 farmers participating in the River Friendly Farms Pilot Test (aggregate average scores on a 1–4 scale, 1 = "low").

	Natural resource conservation	Nutrient management	Pest management	Livestock management
High score	3.5	4.0	3.6	4.0
Low score	2.7	2.3	2.3	2.4
Average score	3.1	3.1	3.0	2.9
Average number of "low"ratings	0.67	0.44	0.53	1.23

age practices, and manure management. This national program, based in Wisconsin, was adapted in many states (see e.g., Hickman et al., 1994) and now includes supplemental programs such as Home-A-Syst (Larson and Johnson, 2000).

In the early 1990s, a consortium of farmers, agency, nonprofit organizations, and university personnel in Ontario, Canada began working to expand Farm-A-Syst to a whole-farm scale. The resulting assessment tool, the Ontario Environmental Farm Plan, uses the Farm-A-Syst rating system (1 = poor, 2 = below average, 3 = adequate or recommended practice, and 4 = exceptional management) and expanded the scorecard to include the whole farm, not just the farmstead. (Ontario Environmental Coalition, 1994). They have met with tremendous success in Ontario securing voluntary participation of farmers in this program, and more than 10 000 farmers participated in the first 5 yr of the program. Many saw the program as a way to prioritize environmental improvements for their farms and stated that the tax rebate and other incentives to participate were less important to them than the environmental incentives (Bidgood, 1995).

A collaboration between Kansas State University and the Kansas Rural Center resulted in the creation of an adapted version of the Ontario Farm Plan model for use in the Midwest. Using feedback from farmers and agricultural extension agents, the authors based many of the "best management practice" questions on the Ontario version, but added new questions relating more to sustainable agriculture, and added farm inventory and goal setting sections (Janke and Nagengast, 1999). Financial planning is also critical to the process of whole-farm planning and is offered as a separate tool through programs like K-State Farm Analysts or Farm Management. A scorecard and action plan are used to summarize the results of the assessment tool, but the scores are not used as an index. Instead, any practice that rates a "poor" on the assessment sheets is prioritized, and next steps are written into an action plan. These action plans are written within the context of the farm goals statement, and then financial analysis is used to generate cost–benefit scenarios for different options. The program is still young, but it has been successfully pilot tested and is in the early stages of state-wide implementation.

Example aggregate data from 21 farms participating in the pilot test are shown in Table 5–1. When diverse farms are summarized together, it is difficult to see much difference in the average scores from the questions in each of the sections. Pulling out only the low scores (1 = poor) is slightly more telling. What we see indicates that among this particular group of producers, more of them had environmental problems related to livestock waste management than they did in the areas of soil conservation, pest, or nutrient management.

Table 5–2. Example question from the Nutrient Management Section, Kansas River Friendly Farm Environmental Assessment Tool: Crop Rotations for Soil Organic Matter Building and Nutrient Buffering (Janke and Nagengast, 2002).

4	3	2	1
Perennial forages or overwintering cover crops grown at least one-third of the time in the rotation. (Wheat [*Triticum aestivum* L.] is considered a crop, not a cover crop in most situations.)	Perennial forages or overwintering cover crops included in the rotation at least once every 10 yr. Diverse crop rotation of at least three to five different types of crops, alternating summer (corn [*Zea mays* L.], milo [*Sorghum bicolor* (L.) Moench], soybean [*Glycine max* (L.) Merr.]) and winter (wheat) crops.	No perennial forages or cover crops, but diverse crop rotation includes at least three to five different types of crops.	Continuous cropping of same crop several years in a row.
Nutrient applications are low to moderate for all crops in the rotation. Fertility is adequate to maintain root and top growth.	Crops that receive high rates of fertilizer are alternated with crops that use residual fertility.	More than one-half the rotation is occupied by crops that receive high levels of nutrients.	The farm is in continuous row crops that receive high levels of nutrients.

It is unclear at this point whether aggregate data will be collected on these farms state-wide for use by policy makers, since the original intent of the tool was for farmer use only, and the more confidential the data is, the better. However, trends with time may help us know at the state level whether we are making progress. For example, if initial scores reveal higher and higher numbers of farmers scoring "1," it may mean that we are reaching new groups of farmers with significant environmental problems on their farms. We could also use the tool to reassess early program participants, to see if their action plan items were completed and whether their individual and overall scores improve. If the number of low scores drop, we have some evidence that environmental conditions on these farms is improving.

As researchers, we can look at this data at the level of the individual questions within each section (Table 5–2) and determine if research is needed in certain areas more than in others. The exercise of writing this assessment tool was helpful to the university community, since we saw more clearly where the gaps in our knowledge and understanding of ecosystem processes were. For example, much disagreement still exists in the research community about how best to build soil. A common comment from farmers who complete the assessment is that they have learned a lot about good environmental management through reading and answering each question in the notebook.

SUMMARY

Whole-farm planning and agroecosystem analysis will require a combination of tools and approaches. From the farmer's perspective, useful tools include maps, record sheets, financial records, and other ways to track production, costs, and returns. In addition, tools that allow a farmer to track resource use (such as water meters on irrigation pumps) and resource (soil and water) quality will be useful. The dynamic of farm planning needs to include goal setting, including all family members and other interested parties. Environmental quality assessment tools such as Farm-A-Syst and the River Friendly Farm Plan allow farmers to find areas of vulnerability on the farm and to make a list of planned projects to correct them.

From a researcher perspective, useful tools include those commonly used in FSR/E, such as farm component diagrams, resource flows (including nutrients), and maps of farms and communities of farms. Looking over the shoulders of farmers and into their record books will also reveal useful information, but the trust of farmers is necessary before they will share private information such as financial records or even environmental assessment worksheets and action plans.

For policy makers, maps of watersheds and farms within watersheds are useful, in addition to knowing specifics about current land use practices on individual fields and farms. Even though we are looking at a specific level in the hierarchical diagram of ecosystems, the farm, we can see that various tools are necessary with differing levels of detail. The level of detail is determined by the audience for that information, and how the information is to be applied.

STUDY QUESTIONS

1. In what ways can whole-farm financial planning complement more detailed assessment of practices and enterprises in decision making?

2. What financial incentives are needed to encourage whole-farm planning?

3. How does the farmer balance financial and other goals in designing a whole-farm system of enterprises?

4. What are the complications that must be faced in developing farm plans that are the result of scale?

5. What are the differences between the goals of a farmer and the goals of a researcher in doing a sustainability index of a farm?

6. How can a farmer connect ecological assessment with economic assessment?

WHOLE-FARM PLANNING EXERCISE

1. Select a real, working farm. This needs to be one with which you are familiar, and/or have a good working relationship with the farm operator/family. Detailed questions will need to be answered as part of this exercise.

2. Find maps of the farm or draw a map of the physical features of the farm.

3. Attempt to diagram the primary enterprises on the farm, similar to the diagram in Fig. 5–2. Show flows coming onto the farm, going off the farm, and how the farm operator/family fits into the diagram.

4. Draw a second diagram, or add more detail to the exercise in Step 3, to show ecosystem processes on this farm, including water flow, sunlight–energy flow, C flow, nutrient (N or P) flow, etc.

5. Complete a whole-farm assessment with the farm operator/family. Use a tool such as Farm-A-Syst, the Ontario Environmental Farm plan, or the Kansas River Friendly Farm plan. Create an action plan after completing the assessment. Discuss what was learned from this exercise with the farm operator/family.

6. Working with an agricultural economist and/or with the farm financial records available through your relationship with the farm operator/family, discuss farm goals with the farm operator/family. What financial information is needed to realize these goals? What role does the environmental assessment play in planning for these goals?

REFERENCES

Arshad, M., and G. Coen. 1992. Characterization of soil quality: Physical and chemical criteria. Am. J. Altern. Agric. 7:25–31.

Bellows, B. (ed.) 1994. SANREM CRSP Indicators of Sustainability Conf. and Worksh. 1–5 Aug. 1994. Center for PVO/University Collaboration in Development, Cullowhee, NC.

Bidgood, M. 1995. A study of actions by Simcoe County environmental farm plan participants. RR1. Barrie, Ontario. Unpublished report to the Ontario Crop and Soil Association, Guelph, ON, Canada.

Binham, S., and A. Savory. 1990. Holistic resource management workbook. Island Press, Washington, DC.

Deutsch, W.G., A.L. Busby, J.L. Orprecio, J.P. Bago, and E.Y. Cequina. 1998. Community-based water quality indicators and public policy in rural Phillippines. p. 1–19. *In* Proc. of the SANREM CRSP/Phillippines 1998 Annual Conference, Malaybala, Phillippines.

Finke, P.A., J. Bouma, and M.R. Hoosbeek (ed.) 1998. Soil and water at different scales. Kluwer Academic Publ., Dordrecht, the Netherlands.

Freyenberger, S., R. Janke and D. Norman. 1997. Indicators of sustainability in whole-farm planning: Literature review. Kansas Sustainable Agric. Ser. 2. Kansas State University, Manhattan.

Granatstein, D., and D. Bezdicek. 1992. The need for a soil quality index: Local and regional perspectives. Am. J. Altern. Agric. 7:12–16.

Hack-ten Broeke, M.J.D., and W.J.M. de Groot. 1998. Evaluation of nitrate leaching risk at site and farm level. *In* P.A. Finke et al. (ed.) Soil and water at different scales. Kluwer Academic Publ., Dordrecht, the Netherlands.

Hickman, J.S., K.L. Herbel, D.H. Rogers, and G.M. Powell. 1994. Farm-a-syst farmstead assessment system. Kansas State Coop. Ext. Serv. Publ. MF-1050. Kansas State Univ., Manhattan.

Hunt, N., and B. Gilkes. 1992. Farm monitoring handbook. 1995. Univ. of Western Australia Press, Perth, Australia.

IOWATER. IOWATER volunteer water monitoring [online]. Available at www.IOWATER.net (verified 10 July 2003). IOWATER, Des Moines, IA.

Janke, R. 2003. Citizen science: Soil and water testing for enhanced natural resource stewardship. Department of Horticulture Fact Sheet S-143. Kansas State Univ., Manhattan.

Janke, R., and S. Freyenberger. 1997. Indicators of sustainability in whole-farm planning: Planning tools. Kansas Sustainable Agric. Ser. 3. Kansas State Univ., Manhattan.

Janke, R., and D. Nagengast. 2002. The river friendly farm environmental assessment tool. S-138. Kansas State Univ., Manhattan.

Karlen, D., N. Eash and P. Unger, 1992. Soil and crop management effects on soil quality indicators. Am. J. Altern. Agric. 7:48–55.

Larson, N., and B. Johnson. 2000. Home-a-syst. Kansas State University Pollution Prevention Institute, Kansas State Univ., Manhattan.

Levins, R. 1996. Financial analysis for sustainable agriculture. Land Stewardship Project, White Bear Lake, MN.

Lightfoot, C., and R. Noble. 2001. Tracking the ecological soundness of farming systems: Instruments and indicators. J. Sustainable Agric. 19:9–29.

Muller, S. 1994. Development of a framework for the derivation of sustainability indicators and application of the framework in the Rio Reventado Watershed in Costa Rica. p. 15–42. *In* B. Bellows (ed.) SANREM CRSP Proc. of the Indicators of Sustainability Conf. and Worksh. 1–5 August 1994. Center for PVO/University Collaboration in Development, Cullowhee, NC.

Ness, J.A. (ed.) 1998. The monitoring tool box. Land Stewardship Project, White Bear Lake, MN.

Ontario Farm Environmental Coalition. 1994. Ontario environmental farm plan. Ontario Farm Environmental Coalition, Toronto, ON, Canada.

Parr, J., R. Papendick, S. Hornick, and R. Meyer. 1992. Soil quality: Attributes and relationship to alternative and sustainable agriculture. Am. J. Altern. Agric. 7:5–11.

Robotham, M.P., and H.J. McArthur, Jr. 2001. Addressing the needs of small-scale farmers in the United States: Suggestions from FSR/E. J. Sustainable Agric. 19:47–64.

Stockle, C.O., R.I. Papendick, K.E. Saxton, G.S. Campbell, and F.K. van Evert. 1994. A framework for evaluating the sustainability of agricultural production systems. Am. J. Altern. Agric. 9:45–50.

Stork, N., and P. Eggleton. 1992. Invertebrates as determinants and indicators of soil quality. Am. J. Altern. Agric. 7:38–47.

Taylor, D., Z. Mohamed, M. Shamsudin, M. Muhayidin, and E. Chiew. 1993. Creating a farmer sustainability index: A Malaysian case study. Am. J. Altern. Agric. 8:175–184.

Visser, S., and D. Parkinson. 1992. Soil biological criteria as indicators of soil quality: Soil microorganisms. Am. J. Altern. Agric. 7:33–37.

6 Multifunctional Economic Analysis

THOMAS L. DOBBS

Economics Department
South Dakota State University
Brookings, South Dakota

It would seem that agriculture in industrialized countries is experiencing the best of times and the worst of times. Productivity per unit of land and, consequently, aggregate food and fiber output have climbed dramatically since World War II. Food is generally affordable relative to average per capita incomes. However, the costs of this abundance are becoming increasingly apparent. Drinking water supplies are becoming contaminated, bird and fish populations have declined, plant and animal biodiversity has been lost, and soil organic matter has declined. Also, agriculture appears increasingly vulnerable to human and animal health scares. Witness the recent outbreaks in Europe of mad cow disease (bovine spongiform encephalopathy, or BSE) and foot and mouth disease. Moreover, hired farm laborers and animal slaughtering house workers often are poorly paid and work in unsafe conditions. In spite of the abundant and cheap food supplies, poverty and malnutrition persist among some groups within the larger, affluent societies of industrialized countries. Persistently *low* crop prices have caused governments to continue, and even increase, large direct payments to farmers to support their incomes. Add to these concerns the increasing economic concentration and vertical integration within both the agricultural supply sector and the agricultural processing and marketing sector that are causing farmers to feel ever more vulnerable to so-called market forces.

This apparent contradiction between agricultural abundance on the one hand and ecological and social breakdown on the other has caused the British government to fold its agricultural and food ministry into a new, combined Department of the Environment, Food, and Rural Affairs. This was a prelude to proposals for radical reform of agriculture in the United Kingdom (UK) that, if enacted, would put much greater emphasis on multifunctionality. Germany has also combined ministries dealing with agriculture, food, and natural resources. In France, the government has recently launched a program of Contrat Territoraile d'Exploitation (land management agreements, or CTEs) as part of its expanded emphasis on multifunctionality. Even the U.S. Congress finally began to openly discuss a more multifunctional approach to agricultural policy in dialogue leading up to the passage of a new Federal farm bill in 2002. Although that new legislation retains and reinforces the historically strong emphasis on support for conventional agricultural commodities, the newly created Conservation Security Program does contain elements of a more multifunctional perspective.

MULTIFUNCTIONALITY

A multifunctionality approach to agricultural policy—or more aptly, *agri-environmental policy*—explicitly acknowledges that agriculture has several functions in addition to producing food and fiber (Dobbs and Pretty, 2001b, p. 2–6). Agriculture also can provide goods and services that are *ecological* and *social* in nature. Ecological or environmental functions include provision of clean water supplies, bird and other wildlife habitat, scenic landscapes, C sequestration (to reduce greenhouse gases and mitigate global warming), and flood protection (by wetlands). Social functions include provision of rural employment and, potentially at least, support for democratic institutions based on Jeffersonian ideals of egalitarian land holding patterns and associated structures of agriculture.

How these multiple functions are expressed is often related to scale. Food production may be primarily for local or regional consumption, or primarily for national or international markets, depending on how agricultural economies are organized. Some ecological services, such as provision of habitat for particular wildlife species, may be primarily local in nature. Clean water supplies may sometimes benefit primarily people outside the local area, but elsewhere within the watershed, namely residents of urban areas. Carbon sequestration is very much a global function of agriculture. Rural employment is primarily local in nature, whereas contributions to Jeffersonian democratic ideals have both regional and national implications.

It is apparent that multifunctional economic analyses must start with particular spatial perspectives. For purposes of this chapter, I will assume three different perspectives: (i) that of the individual farm, where decisions in market economies about production practices and systems have traditionally been made; (ii) that of the local or regional level, ranging from the community surrounding a rural town up to watershed, state, and multistate areas representing particular types of agricultural ecosystems; and (iii) that of national and international economies, where effects of agricultural systems on food prices, human and animal health, and climate change are felt, and where public policies are made that powerfully impact the structure and directions of agriculture.

ECONOMIC TREATMENT OF MULTIFUNCTIONALITY

Economists have been using some economic analysis methods for many years that directly relate to agriculture's multifunctionality. The field of natural resource economics (earlier known as land economics) has roots in late 18th and early 19th century writings of the British classical economists Adam Smith, David Ricardo, and Thomas Malthus. Land economics became an important area of inquiry in the USA in the late 19th and early 20th centuries, drawing initially on the German historical school of economics. The University of Wisconsin was an early and prominent leading institution in land economics research and teaching. The economics of conservation took on prominence in the 1950s at the University of California, under the leadership of Ciriacy-Wantrup, and comprehensive work on a broad set of natural resource economics concerns began to take place under the leadership and sponsorship of Resources for the Future, established in Washington, DC in 1952. The body of work known as modern welfare economics began to be systematically included in natural resources research and policy analysis. This led to many published works about, and a deeper understanding of, "externality" and "public goods" dimensions of agriculture in the 1950s, 1960s, and 1970s (Castle et al., 1981). Resource economists devoted substantial attention to the valuation of "nonmarket" environmental goods during the last three decades of the 20th century.

While the concepts of externalities and public goods dominated mainstream natural resource economics work from the 1960s through the 1990s, a related but more radical form of economic thinking about agricultural and other natural resources arose during the 1980s: "ecological economics" (Costanza et al., 1997, p. 49). This emerged out of a combination of ecology and economics, much as the discipline of agricultural economics had emerged out of a combination of agronomy and economics around the beginning of the 20th century (Dobbs, 1987). Beginning in the mid 1960s, the writings of University of Colorado economist Kenneth Boulding were instrumental in helping to bring together the thinking of ecology and economics, disciplines which had been pursued largely apart from each other during most of the 20th century (Costanza et al., 1997, p. 51). Ecological economics takes a more holistic perspective than does natural resource economics, and, drawing on its ecology roots, explicitly considers the interconnectedness of the various components of natural resource systems. Ecological economics also poses fundamental questions about the ecological sustainability of different economic systems, whereas natural resource economics tends to take market-based economic systems as given and considers how the systems might be refined to produce more socially desirable ecological results. In fairness, ecological economics draws heavily on the body of theory that was built up over a century in "land" and "natural resource" economics. However, ecological economics opens the door to a wider range of policy considerations than, traditionally applied, does natural resource economics.

In the sections to follow, I will draw on both natural resource economics and ecological economics, although I will not stress distinctions between the two. First, I will discuss applications of multifunctional economic analysis at the farm level. Then, I will discuss applications at the local or regional level, drawing heavily on natural resource economics. In the last section, which deals with multifunctional economic analysis at the national and international level, I draw on both natural resource economics and ecological economics.

ANALYSES AT THE FARM LEVEL

Following Crews et al. (1991), I usually define agricultural sustainability in ecological terms. Once ecological sustainability objectives have been specified at the farm level, economics enters the picture. Farming systems thought to satisfy ecological sustainability objectives must be economically attractive to farmers if they are to be voluntarily adopted and continued.

Farmer Incentives

Goals to maintain or increase profits are central to virtually all analyses of farmers' incentives to adopt new or different technologies or systems. However, it is well recognized that goals related to risk also play an important role in farmers' decisions. The risk-averse behavior that farmers (like other people) often exhibit can have substantial implications for agri-environmental policy. Oglethorpe (1995), for example, demonstrated that greater variability in the market for agricultural goods can be more conducive to farmer-adoption of less intensive practices than would be the case with more stable markets. One implication of this is that if government income protection policies take most of the risk out of intensive, specialized farming, the costs of utilizing agri-environmental policies to induce farmers to voluntarily adopt less intensive, more ecologically sustainable farming systems can be high.

Many farmers obviously have other goals as well, including having adequate leisure and family time and having a sense of independence. A set of goals with special relevance to this chapter, however, relates to stewardship of natural resources. Some farmers base their farm stewardship decisions on ethical grounds to a much greater extent than do others, where ethical is meant to include reasons other than pure self interest (Colman, 1994). Thus, it is logical to posit a stewardship goal when examining farmer incentives to adopt ecologically sustainable systems. This goal can be interrelated with goals to have social standing in the local community, when and where the public values sound stewardship.

Other considerations certainly enter into farmers' decisions about adoption of sustainable practices, including the ability to maintain flexibility and not be tied down by bureaucracy. While these considerations should not be ignored, the primary focus in this chapter—following Dobbs and Pretty (2001a)—is on the following three sets of farmers' goals:

- To maintain or increase net income (profits) from farming
- To avoid *excessive* risk with the income-generating activities of the farm
- To maintain sound stewardship of the farm's natural resources

These goals are the ones thought to be most directly relevant to farmers' decisions about adoption and continued use of ecologically sustainable farming practices.

Economic Measurement

Sometimes economic measurements to determine farmers' incentives to change to more environmentally sound farming *practices* can be carried out with farm management budgeting on a single crop or enterprise. Changes in tillage practices or timing of fertilizer applications are examples where such an approach might suffice. Moves to fundamentally different farming *systems*, however, require measurement of economic changes across the entire farm or at least a major portion of the farm. A change from a chemical-intensive corn (*Zea mays* L.)–soybean [*Glycine max* (L.) Merr.] crop system with no livestock to a farming system that integrates small grains, forage legumes, and one or two livestock enterprises with reduced-chemical corn and soybean would call for a whole-farm analysis that accounts for the numerous interacting changes that impact profits, risk, and natural resource stewardship. An interrelated set of changes in crop rotations and fertility and pest control methods would necessitate, at least, a whole-rotation analysis. Side-by-side economic comparisons of profits and risks of individual crops—say, comparison of corn in the farmer's current system to corn in the new or proposed system—would have little meaning. It is only the net, overall effect on profits, risk, and stewardship that has meaning for such system changes.

Madden and Dobbs (1990) discussed some of the procedures and considerations in carrying out whole-farm and other systems analyses. There often seem to be conflicting findings among various economic analyses comparing conventional and more ecologically sustainable farming systems. Among the reasons that underlie *apparent* conflicts in findings from different studies are the following questions (Dobbs, 1994b):

- Are short-run or long-run measures of profitability used?
- Are federal farm program provisions accounted for in the whole-farm models or are they ignored (or greatly simplified)?
- Is family labor included as a cost in the enterprise budgets?

The approaches taken sometimes depend on the question(s) being asked. In our farming systems studies at South Dakota State University (SDSU), we generally have asked: "Is there adequate incentive over the long run for farmers in particular agroclimatic areas to adopt organic farming systems or other systems with dramatically lower chemical-input needs?" Consequently, we have attempted to include all costs related to conventional and alternative systems, including machinery depreciation costs and family labor costs. Over the long run, machinery must be replaced, and even family labor has opportunity cost. We also have generally included farm program payments and related provisions (e.g., land set-aside requirements when they existed), since payments and restrictions in some form seem to be permanent fixtures of agriculture in industrialized countries. However, since the level and forms of payments and related provisions are constantly changing, we often conduct sensitivity analyses to observe how economic comparisons are affected by possible changes.

Applications

Under the leadership of James Smolik, researchers at SDSU carried out two sets of studies at the Northeast Research Station (near Watertown, SD) from 1985 through 1992, comparing selected alternative (organic), conventional, and reduced tillage crop rotation systems. Descriptions of the systems and the agronomic and economic methods of analysis are presented in Smolik et al. (1995). This was a whole-system analysis of the crops portion of hypothetical farms using the different systems. Possible changes in livestock that might result from changes in crop systems were not explicitly considered. We assumed that markets would be available for hay produced by one of the alternative (organic) systems. A summary of the profitability comparisons from 1986 (omitting 1985, the establishment year) through 1992 is shown here as Table 6–1.

Several measures of net income, averaged for the 7-yr study period for each system, are shown in this table. The first measure, shown in the third column of data,

Table 6–1. Average economic results in the Farming Systems Studies (1986–1992).†

System‡	Direct costs other than labor	Gross income	Net income over: All costs except land, labor, and management	Net income over: All costs except land and management	Net income over: All costs except management	Whole farm net income over all costs except management§
	$ ha^{-1}					$
Farming Systems Study I						
1. Alternative (oat [*Avena sativa* L.]–alfalfa–soybean–corn)	113	383	188	158	93	20 088
2. Conventional (corn–soybean–spring wheat [*Triticum aestivum* L.])	155	378	145	123	58	12 528
3. Ridge till (corn–soybean–spring wheat)	173	348	103	80	15	3 240
Farming Systems Study II						
1. Alternative (oats–clover [*Trifolium* spp.]–soybean–spring wheat)	75	253	118	95	30	6 480
2. Conventional (soybean–spring wheat–barley [*Hordeum vulgare* L.])	120	318	123	98	33	7 128
3. Minimum till (soybean–spring wheat–barley)	148	290	73	50	–15	–3 240

† Adapted from Dobbs (1993, p. 51).
‡ Crops are shown in the order in which they occur in each rotation.
§ For farm with 216 tillable hectares.

includes a deduction for all costs (including items like machinery depreciation and interest) except land, labor, and management. The next measure (in the fourth column) includes all costs in the first measure plus a charge for labor used for crop production. A land charge is included in the final measure (in the fifth column); this final measure is referred to as "net income over all costs except management." The land charge is the same for all systems. This final measure constitutes what is often referred to as "pure profit" or "return to management for planning and risk taking." The same measure is shown in the last column of Table 6–1, except that it is on a whole-farm (216 tillable hectares) basis.

Just for illustrative purposes, see the net income over all costs except management data in the last two columns. The alternative system was the most profitable in Study I. Net income per hectare was very similar for the alternative and conventional systems in Study II, and net income for the minimum till system in that study actually was negative, on average—meaning gross returns fell short of covering all costs for that system (Dobbs, 1993).

Management costs here refer to opportunity costs of the farmer's time spent in planning, observing crops, making production-related decisions, and marketing. They were not included simply because of the difficulty in objectively measuring differences among systems in those costs. Overall, management costs probably are somewhat higher for the types of alternative systems we analyzed than for the other systems. Therefore, a net income measure that also factored in management costs probably would have slightly reduced the profit advantage of the alternative system in Study I, and it probably would have caused the slight advantage of the conventional system in Study II to increase (relative to the alternative system).

The federal farm program during the study period affected the net income of the various systems differently, with the alternative system receiving an average of \$22.50 ha^{-1} yr^{-1} less in government payments than the other systems in Study I. In Study II, the alternative system received an average of \$10 ha^{-1} less than the other systems. (Dobbs, 1994a).

We also examined the relative attractiveness of these systems to farmers in terms of risk. One measure of the risk associated with different farming systems is the variability in profits over time, as measured by the standard deviation. Net income over all costs except management was much less variable for the alternative system in Study I, for example, than for the conventional and ridge till systems. The standard deviation for the alternative system was \$41.50 ha^{-1}, compared with \$78.25 ha^{-1} for the conventional system and \$74 ha^{-1} for the ridge till system. Also, the alternative system never had negative net returns; this downside risk protection is an important consideration for most farmers (Smolik et al., 1995).

What about measures of the possible implications of these different systems for farmers' stewardship goals? It is difficult, if not impossible, to quantify all the relevant dimensions of stewardship. The research team did examine possible soil erosion and water quality implications. The alternative and reduced till systems were the only systems that consistently met or exceeded Federal farm program conservation compliance regulations for residue cover on highly erodible land. Therefore, it was concluded that both types of systems should adequately address soil erosion concerns and may help limit N and P runoff. Nitrate-N levels in Study I were measured by soil cores in the last year, and were found to be about two to three times higher in conventional and ridge till systems than in the alternative system. Higher NO_3–N levels may not necessarily indicate an environmental problem if groundwater is not vulnerable, or if the N is in the upper layers of the soil profile, where crops can readily use it. However, most of the NO_3–N in the conventional and ridge till systems was below 0.61 m (2 feet), suggesting the potential for groundwater pollution is greater for those systems than for the alternative system (Smolik et al., 1995).

The economic results cited here for our studies at SDSU's Northeast Research Station were based on the assumption that no organic price premiums were received for production from the alternative (organic) systems, even though farmers using those systems would have been eligible for organic certification and any available price premiums after 3 yr. We made this assumption to be conservative with respect to organic market expectations, in case substantial expansions in organic acreages wipe out or greatly diminish price premiums. In a companion study comparing an actual operating organic farm in east-central South Dakota and a neighboring conventional farm, we did include organic price premiums for years in which we had appropriate data. In that study—covering 1985 through 1992—the organic farm was less profitable than the conventional farm most years, when organic price premiums were not factored in. For the 4 yr, from 1989 to 1992, for which we had information on price premiums the organic farmer received on some of his production, including those premiums added an average of \$27.50 ha^{-1} to net income over all costs except management. This was enough to narrow but by no means close the net returns gap between the organic and the conventional farm during that period (Dobbs and Smolik, 1996). Organic price premiums generally were much higher in the late 1990s (Bertramsen and Dobbs, 2001, 2002), so the results might have been different had the study been continued for several more years.

ANALYSES AT THE LOCAL AND REGIONAL LEVEL

Multifunctional economic analysis at the local or regional level takes account of certain external effects of agricultural production systems that are not necessarily considered by farmers in their management decisions. The natural resource economics literature dealing with market failure provides the theoretical basis for this analysis.

Market Failure

Externalities consist of spillover effects from agricultural production, either positive or negative, that are uncompensated (or not accounted for in farmers' costs, if negative) in unregulated markets. Since externality values are not reflected in market transactions, markets often fail to produce the socially desired amounts of primary agricultural commodities and the externalities associated with those commodities. In the case of negative externalities, such as NO_3 contamination of local groundwater by intensive corn production, "too much" corn is produced and "too much" polluted water (and "too little" clean water) is produced. Farming systems that incorporate forage or green manure legumes and small grain crops in rotations with row crops provide a biologically diverse habitat for many birds and beneficial insects. This can be thought of as a positive external effect, in part, of forage and green manure legumes that is uncompensated in the market. Consequently, "too little" of such legumes is grown, relative to, say, corn and soybean (Prato, 1998, p. 91–104; Zilberman and Marra, 199, p. 221–234).

Market failure is based on technological (or physical) externalities, rather than on pecuniary (or price-effect) externalities. Pecuniary externalities result, for example, when individuals or firms purchase or sell large enough quantities of a good or service to affect price levels. The change in price levels affects people who are not directly involved in the original transactions, but who now face higher or lower prices as a result of those original transactions. These pecuniary externalities help some groups and hurt others, but they do not necessarily constitute a failure of the market economy (Davis and Kamien, 1972). In contrast, technological externalities, such as the off-site effects of soil erosion, often do result in market failure. It is technological externalities that are commonly simply termed "externalities" in most of the current environmental literature (see Common, 1995; Knutson et al., 1998).

Many of the positive externalities relevant to multifunctional analysis at the local or regional level have characteristics of what economists call public goods. These are goods for which no one (practically speaking) can be excluded from enjoying, and use by one individual does not diminish the availability of the good for use by other individuals. Examples include certain types of wildlife for observation and scenic rural landscapes (Mullarkey et al., 2001). Of course, beyond some point of crowding, some such amenities could lose the characteristic of noncompetitiveness in use.

An important consideration in multifunctional economic analysis is the extent of jointness in production (Cahill, 2001). A true technological externality implies some jointness, say between dairy production in Wisconsin and a scenic landscape dotted with big white barns. In that example, the end of dairy production in an area could spell the eventual deterioration and disappearance of the scenic barns. On the other hand, more cows or more milk production per cow will not necessarily lead to more barns and improved scenery. Thus, there is some jointness, but it is not absolutely rigid. How best to save the white barns depends in part on this degree of jointness. One might let dairy production disappear entirely in particular areas, yet pay farmers for maintaining the barns (i.e., for the scenic function), but would barns without any dairy cows at all around them or in nearby pastures provide quite the same scenery? The point here is that analyses of policy alternatives for encouraging the beneficial multiple functions of agriculture must consider the nature and degree of jointness between commodity and noncommodity outputs (Cahill, 2001).

I will return to the topic of policy alternatives in the section on multifunctionality analysis at the national and international levels. At this point, however, I turn to economic measurement of noncommodity outputs of agriculture.

Economic Measurement

By their very nature, many of the noncommodity outputs of agriculture are difficult to quantify, especially in monetary terms. Consequently, nonmonetary measures of benefits or costs frequently are relied upon heavily. Examples might be bird species restored, reductions in soil erosion in terms of tonnes of soil per hectare, kilometers of public footpaths provided through scenic farm landscape, or hectares of native grassland preserved.

Economists have devoted a great deal of effort in the last several decades to developing techniques for placing monetary values on nonmarket goods. The economic theory upon which these techniques are based involves concepts of willingness to pay and willingness to accept compensation. Both indirect and direct methods of measurement have been used. An indirect method that has been much used for certain kinds of recreational experiences is the travel cost approach. In this approach, travel distances and costs of participants in a recreational activity are used to estimate demand, or willingness and ability to pay, functions for the activity. The contingent valuation method (CVM) is the most common direct valuation method. This method relies on surveys of sample households in which responses are elicited about willingness to pay or to accept compensation (Prato, 1998, p. 305–329).

Applications

A study of the relationships between farming practices and systems and water quality in the Sioux River drainage of eastern South Dakota illustrates quantitative, but nonmonetary measurement of an environmental externality. Dobbs and Bischoff (1996) estimated net returns to land and management for different practices and crop rotation systems on four farms in the study area, as well as the probable NO_3 leaching to groundwater with each of the same combinations of farm practices and systems. A model called the Nitrate Leaching and Economic Analysis Package (NLEAP) was used to estimate NO_3 leaching. The relationships measured in this study for one of the case study farms are shown in Fig. 6–1. This was an irrigated farm on which continuous corn was grown on 29 ha (the Before scenario in Fig. 6–1). The other scenarios represent various changes that might have been possible, with financial assistance at the time from an agrienvironmental program called the Water Quality Incentives Program (WQIP). Estimated net returns increased by \$44 ha^{-1} (29%) where the WQIP involved eliminating dry preplant inorganic fertilizer (i.e., moving from the Before to the After scenario); however, the NO_3 leaching did not change much. Profitability was 8% greater when the alternative practice of splitting N applications also was added to the After scenario. Changes that involved alternative crop rotations lowered net returns—to \$187 ha^{-1} for a corn–soybean rotation and to \$135 ha^{-1} for a 6-yr rotation involving alfalfa (*Medicago sativa* L.), corn, and soybean—compared with \$203 ha^{-1} in the After scenario and \$220 ha^{-1} in the splitting N scenario. Environmental results for splitting N applications indicated a 7% decrease in the amount of NO_3 leached compared with the After scenario. Changing rotations showed greater decreases in NO_3 leaching, however. Estimated NO_3 leached went from 41 kg ha^{-1} in the After scenario to 30 kg ha^{-1} in the corn–soybean rotation system and to 29 kg ha^{-1} in the 6-yr crop rotation system (Fig. 6–1).

Our research comparing organic systems with other farming systems in northeastern South Dakota, discussed in the section above on analyses at the farm level, can be used to illustrate nonmonetary measurement of a social function. There is a great deal of interest in how different farming systems affect farm size and the resulting structure of agriculture. Many factors affect farm size, including the farm's ability to generate adequate levels of family income. Therefore, we analyzed the potential farm size implications of the labor and management returns associated with different systems included in that research. In Study I, where systems included substantial amounts of corn and soybean and where labor intensiveness was greater for the organic system, a conventional farm would have had to be nearly twice as large as an organic farm to generate $40 000 in net returns to family labor and management. On the other hand, in Study II, where systems consisted primarily of small grains, labor intensiveness was lower for the organic system and there was little difference between the organic and conventional systems in returns per hour of labor and management. Therefore, the implied difference in farm size was much less in Study II (Smolik et al., 1995).

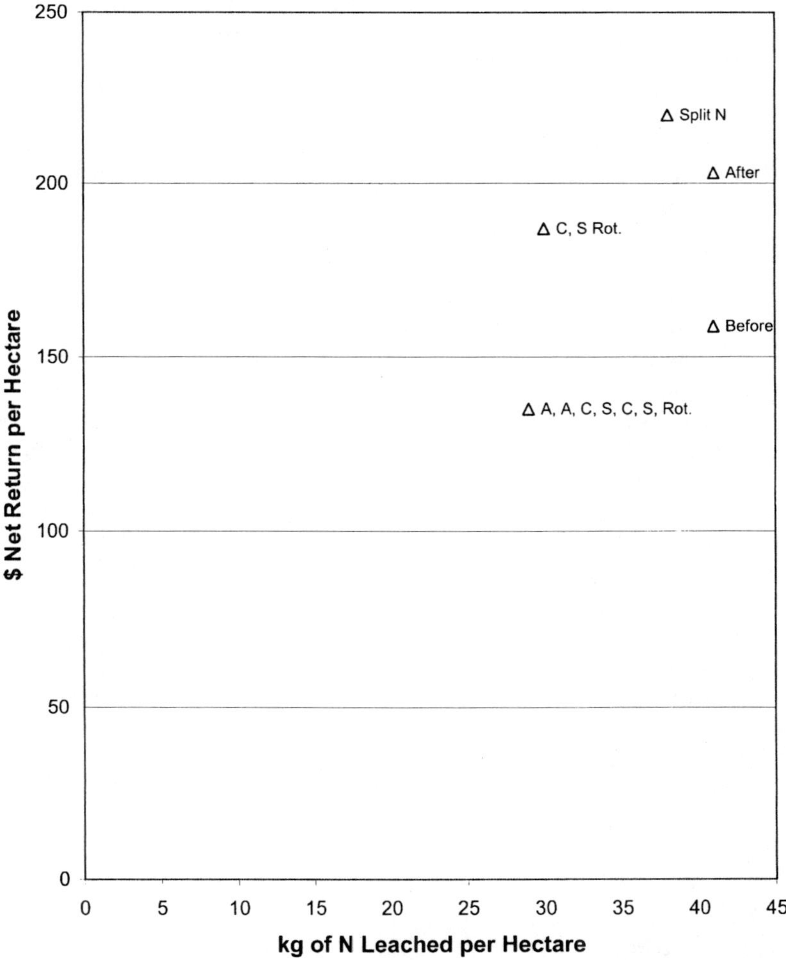

Fig. 6–1. Profitability/N leaching relationships: Case Farm 4 (typical year). Adapted from Dobbs and Bischoff (1996).

Freyenberger et al. (2001) used three indicators of sustainability to compare a group of conventional farms in Kansas with a group of alternative farms in the same state. They compared the percentages of gross farm income that were accounted for by types of energy and machinery expense that tend to be associated with environmental problems or that involve few gains to local economies because most of the dollars leave the local community. They also compared the percentages of gross income made up of wages paid to hired labor—larger percentages were preferred because of the presumed benefits of local job creation. In addition, they used a feed balance indicator of the "degree to which livestock feed production and feed consumption are in balance on the farm" (Freyenberger et al., 2001, p. 33). Alternative farms were represented by a selected sample associated with the Kansas Rural Center (KRC), a sustainable agriculture advocacy group. The KRC farms were found to average

> ...lower indicator values for energy and machinery use, higher values for creation of local jobs, and lower values for feed balance, all characteristics desired for farms being managed in a sustainable manner.
>
> (Freyenberger et al., 2001, p. 33)

Economists in Iowa recently estimated willingness to pay for either prevention of further deterioration of water quality or for improved water quality (Azevedo et al., 2001). This was part of a comprehensive analysis of alternatives to maintain and improve water quality in Clear Lake, Iowa. Monetary estimates were made of the value both visitors and local residents placed on preservation and/or restoration of the lake. Estimates revealed a high willingness to pay to avoid further deterioration of the lake, a willingness to pay only moderate amounts for a low quality improvement, and a willingness to pay substantial amounts for a significant quality improvement to lake conditions.

Economists in the UK have conducted extensive analyses of agri-environmental schemes carried out there since the mid 1980s. Hanley et al. (1999) recently summarized a review by Stewart et al. (1997) of the major cost–benefit analyses of those schemes. Ten of the twelve schemes covered in this review were part of the Environmentally Sensitive Areas (ESA) program. When the ESA program was launched in 1986, it was the first agri-environmental program in the European Union. The CVM method was used in most of the studies to estimate monetary benefits of the schemes. In many of the evaluations, a range of benefit estimates was presented. Benefit estimates at the upper ends of those ranges exceeded costs for each of the ESA schemes, sometimes by many times the costs. Hanley et al. (1999) discussed a number of problems associated with such evaluations, as did Whitby (2000, p. 324–325), who noted that most evaluations of UK agri-environmental schemes have not actually been able to value benefits at the margin. In other words, even if the ESA scheme in one area has produced more environmental benefits than costs, that does not necessarily imply that expanding the scheme or adding similar schemes elsewhere will produce greater benefits than costs.

Although much researched by economists, monetary measurement of non-market outputs of agriculture remains a difficult and challenging task. It is not something to be undertaken lightly or by analysts not fully versed in the various available methods and their strengths and shortcomings.

ANALYSES AT THE NATIONAL AND INTERNATIONAL LEVEL

In many ways, multifunctional economic analyses at the national and international level are similar to ones at the local or regional level, especially in the sense that externality effects constitute a central consideration. However, analyses of policy alternatives for dealing with agriculture's multiple functions generally need to be national and international in scope. While local and regional policies are important, and certainly can make a difference, they often can be undermined by contrary policies at the national level or in other countries.

Externality Effects on a Broader Scale

Analyses of agricultural externality effects at the national scale, or other scales broader than local or regional levels, call into question the sustainability of modern, conventional agricultural systems. One of the most recent and comprehensive such analyses was conducted in the UK under the auspices of the University of Essex's Centre for Environment and Society (Pretty et al., 2000). This study, carried out by a multidisciplinary team of researchers based at several different institutions, covered a number of different types of external environmental and health costs of modern agriculture in the UK. Seven broad categories of such costs were examined: (i) damage to natural capital—water; (ii) damage to natural capital—air; (iii) damage to natural capital—soil; (iv) damage to natural capital—biodiversity and landscape; (v) damage to human health—pesticides; (vi) damage to human health—nitrate; and (vii) damage to human health—microorganisms and other disease agents.

Aggregating the results across all of these categories, the researchers found total external costs of UK agriculture in 1996 to be the equivalent of roughly \$3.6 billion, or \$318 ha^{-1} of arable land and permanent pasture (1 U.S. dollar = 0.65 British pound in July 2002). These estimated externalities are equivalent to 89% of average net farm income and 13% of average gross farm income in the UK during the 1990s. Of course, there is a great deal of uncertainty associated with any such estimates. Therefore, the researchers found the estimated range of annual external costs for the period 1990 to 1996 (converted here to U.S. dollars) to be \$1.8 to \$6 billion (Pretty et al., 2000).

As large as these estimates are, they are based only on externalities that produce financial costs. Therefore, the authors state that they likely underestimated the total negative impacts of modern agriculture (Pretty et al., 2000). Regardless of the exact values, the magnitudes of external costs are alarmingly high, as they no doubt are in many other countries where agriculture has become highly industrialized.

Analysis of Policy Alternatives

I turn now, in the final section of this chapter, to economic analysis of policy approaches for addressing externality and related multifunctionality concerns. Natural resource economics and, more recently, ecological economics have contributed greatly to our understanding of how best to encourage the positive dimensions and discourage the negative dimensions of agriculture's multifunctionality. Broadly speaking, the alternative approaches can be classified as either regulatory or incentive based.

Regulatory approaches generally set limits that cannot be exceeded for negative externalities or other adverse effects of agriculture; or, sometimes they specify required (minimum) levels of positive action or provision of nonmarket goods. Conservation compliance provisions of U.S. federal farm policy, dating back to 1985, require minimum ground cover on highly erodible soils as a condition for receiving commodity support payments. Although farmers could forego such payments and thereby not be bound by these provisions, payments have been so substantial most years that the provisions effectively constitute regulations. In comparing the relative effectiveness of regulatory and incentive-based approaches for dealing with environmental concerns, Costanza et al. (1997) list the following advantages and disadvantages of regulatory approaches:

Advantages

- Simplicity, familiarity, and acceptance.
- Historical U.S. reliance upon legislative regulation in order to deal with perceived problems.
- Acceptance by emitters and interest groups.
- Long-term incorporation into the legal system.

(Costanza et al., 1997, p. 196)

Disadvantages

- Effective regulation requires a level of technical and proprietary information that is seldom available to regulators.
- Successful enforcement of regulation requires high monitoring and enforcement costs.
- The costly bureaucracies associated with regulation result in high expenditure per unit of pollution reduction.
- Environmental regulations are easily evaded or avoided.
- The lack of strong incentives to reduce pollution below the mandated level reduces motivation for technological advance and for preventing pollution before it is generated.
- Polluters are permitted to ignore the costs their actions impose upon society *at the time decisions are made*.

(Costanza et al., 1997, p. 196–197)

Incentive-based approaches attempt to take advantage of market forces as much as possible, in order to achieve least-cost solutions to environmental problems. Examples of such approaches are pollution-based taxes or fines and incentive payments based on the extent of positive environmental performance. Potential advantages and disadvantages of incentive-based approaches, in comparison to regulatory approaches, include the following:

Advantages

- They have the ethical advantage of consistency with the. . "polluter pays" principle.
- They raise public revenues.

- They pass the cost of pollution control along to the consumer of pollution-intensive products…
- They provide polluters with economic incentives to prevent pollution…
- Marketable permits do not require that regulators have the level of technical proprietary information required for efficient regulation.
- They can provide incentives for shifting the burden of monitoring from the government to the polluter.
- They offer profitable opportunities for industry to undertake development projects for improvements in pollution abatement technology.
- They can shift the incidence of tax burdens away from socially desirable objectives (incomes and jobs) toward reducing socially undesirable phenomena (pollution).

(Costanza et al., 1997, p. 205)

Disadvantages

Incentive-based approaches based on market theory do not directly address the issues of:

- sustainable scale;
- income distribution, or equity, and therefore of unequal access to environmental protection among individuals, nations, regions, and generations;
- limitations of scientific information and of knowledge by individuals on the ability to make wise choices; and
- additionally the market failures that would need correction in order to make markets work for environmental quality are numerous and perverse. They include externalities, excessive time discounting, common property resources, open-access resources, public goods, and noncompetitive markets.

(Costanza et al., 1997, p. 205)

Up to the present time, policies for agriculture in the USA and European Union countries have emphasized incentive-based approaches. Moreover, the incentives used have generally been positive, such as cost-share agri-environmental schemes, rather than negative, such as taxes on polluting practices or inputs (though such have been used some). Reasons for heavy reliance on the incentive-based approach in agriculture include the first four disadvantages (required level of technical and proprietary information seldom available to regulators, high monitoring and enforcement costs, costly bureaucracies resulting in high expenditure per unit of pollution reduction, and ease of evading regulations) of the regulatory approach listed above and the fourth item (economic incentive to prevent pollution) in the above list of advantages of incentive-based approaches. Dobbs and Pretty (2001a,b) examined a number of issues associated with continued reliance on and expansion of incentive-based agri-environmental schemes to provide positive nonmarket goods and discourage negative environmental externalities from agriculture, including the following:

- To what extent are continued government financial subsidies for the commodity production function of agriculture compatible with agriculture's environmental stewardship functions?
- How can governments strike an appropriate balance between agri-environmental stewardship payments and regulatory approaches?
- Can agri-environmental programs simultaneously promote social and stewardship functions of agriculture?

- Under what conditions are stewardship payment schemes likely to be compatible with World Trade Organization rules?
- Should farmers who are already practicing good environmental stewardship be eligible for agri-environmental incentive payments?

It remains to be seen how long citizens in North America and Western Europe will continue to support environmental approaches that rely primarily on incentive payments to farmers. The UK government has already shown some shift in its approach to reducing NO_3 contamination from agriculture by beginning to phase out its incentive-based Nitrate Sensitive Areas scheme and placing primary reliance on the regulatory Nitrate Vulnerable Zones scheme (Dobbs and Pretty, 2001a). As Western governments place increased emphasis on agriculture's multifunctionality in public policies, the challenge will be to strike an appropriate balance between policy options in three areas (Pretty et al., 2001): (i) environmental taxes, (ii) production subsidy and agri-environmental incentive reform, and (iii) institutional and participatory mechanisms. Approaches in this third area rest on the knowledge that if agriculture is truly to be sustainable in all of its important functions, farmers eventually must take ownership of sustainable approaches, and not just be forced to continue certain approaches by government carrots and sticks. Institutions must exist that foster and support ongoing learning and adjustment, since there never will be a simple set of formulas for agricultural sustainability.

STUDY QUESTIONS

1. What are the public policy implications of viewing agriculture in a multifunctionality context, rather than in the context only of food and fiber production?

2. Why does the existence of externalities in agriculture sometimes call for public policy interventions? Compare the nature of externalities at the local level to externalities at the national level, and how policy responses might differ accordingly.
3. Discuss alternative approaches—and advantages and disadvantages of those approaches—for measuring and valuing externality effects of agriculture. Include an explanation and discussion of the Contingent Valuation method.

4. Using specific agricultural externality examples, discuss advantages and disadvantages of using (a) regulatory and (b) incentive-based policy approaches to address the problem.

5. What might be some reasons for using each of the different measures of net income shown in Table 6–1 of this chapter? Clearly explain the meaning of each measure.

REFERENCES

Azevedo, C., J.P. Herriges, and C.L. Kling. 2001. Willingness to pay for Clear Lake cleanup. Iowa Ag Rev. 7(3):4–5. 8.

Bertramsen, S.K., and T.L. Dobbs. 2001. Comparison of prices for 'organic' and 'conventional' grains and soybeans in the Northern Great Plains and Upper Midwest: 1995 to 2000. Economics Pamphlet 2001-1. South Dakota State Univ., Brookings, SD.

Bertamsen, S.K., and T.L. Dobbs. 2002. An update on prices of organic crops in comparison to conventional crops. Economics Commentator 426(February 22):1–4.

Cahill, C. 2001. The multifunctionality of agriculture: What does it mean? EuroChoices. Spring, p. 36–40.

Castle, E.N., M.M. Kelso, J.B. Stevens, and H.H. Stoevener. 1981. Natural resource economics, 1946–75. p. 393–500. *In* L.R. Martin (ed.) A survey of agricultural economics literature. Vol. 3: Economics of welfare, rural development, and natural resources in agriculture, 1940s to 1970s. Univ. of Minnesota Press, Minneapolis.

Colman, D. 1994. Ethics and externalities: Agricultural stewardship and other behavior: Presidential address. J. Agric. Econ. 45:299–311.

Common, M. 1995. Sustainability and policy. Cambridge University Press, Cambridge, MA.

Costanza, R., J. Cumberland, H. Daly, R. Goodland, and R. Norgaard. 1997. An introduction to ecological economics. St. Lucie Press (for International Society for Ecological Economics), Boca Raton, FL.

Crews, T.E., C.L. Mohler, and A.G. Power. 1991. Energetics and ecosystem integrity: The defining principles of sustainable agriculture. Am. J. Altern. Agric. 6(3):146–149.

Davis, O., and M. Kamien. 1972. Externalities, information, and alternative collective action. p. 69–87. *In* R. Dorfman and N. Dorfman (ed.) Economics of the environment: Selected readings. W.W. Norton & Company, New York.

Dobbs, T.L. 1993. Economic relationships. p. 49–53. *In* J.D. Smolik (ed.) Agronomic, economic, and ecological relationships in alternative (organic), conventional, and reduced-till farming systems. Agric. Exp. Stn. Bull. 718. South Dakota State Univ., Brookings.

Dobbs, T.L. 1994a. Organic, conventional, and reduced till farming systems: Profitability in the Northern Great Plains. Choices 9(2):31–31.

Dobbs, T.L. 1994b. Profitability comparisons: Are emerging results conflicting or are they beginning to form patterns? p. 33–39. *In* Sustainable agriculture: Conceptual and methodological issues. Proc. of an Organized Symposium at Annual Conference of the American Agricultural Economics Association. Cooperative Agricultural Research Program, Tennessee State Univ., Nashville, TN.

Dobbs, T.L. 1987. Toward more effective involvement of agricultural economists in multidisciplinary research and extension programs. West. J. Agric. Econ. 12:8–16.

Dobbs, T.L., and J.H. Bischoff. 1996. Potential for cost-share policies to improve groundwater quality without reducing farm profits. Economics Staff Paper 96-4. South Dakota State Univ., Brookings, SD.

Dobbs, T.L., and J.P. Pretty. 2001a. Future directions for joint agricultural-environmental policies: Implications of the United Kingdom experience for Europe and the United States. South Dakota State Univ. Econ. Res. Rep. 2001-1 and Univ. of Essex Centre for Environ. and Society Occasional Paper 2001-5. South Dakota State Univ., Brookings, and Univ. of Essex, Colchester, England.

Dobbs, T.L., and J.P. Pretty. 2001b. The United Kingdom's experience with agri-environmental stewardship schemes: Lessons and issues for the United States and Europe. South Dakota State Univ. Econ. Staff Paper 2001-1 and Univ. of Essex Centre for Environ. and Society Occasional Paper 2001-1. South Dakota State Univ., Brookings, and Univ. of Essex, Colchester, England.

Dobbs, T.L., and J.D. Smolik. 1996. Productivity and profitability of conventional and alternative farming systems: A long-term on-farm paired comparison. J. Sustainable Agric. 9:63–79.

Freyenberger, S., R. Levins, D. Norman, and D. Rumsey. 2001. Beyond profitability: Using economic indicators to measure farm sustainability. Am. J. Altern. Agric. 16:31–34.

Hanley, N., M. Whitby, and I. Simpson. 1999. Assessing the success of agri-environmental policy in the UK. Land Use Policy 16:67–80.

Knutson, R., J. Penn, and B. Flinchbaugh. 1998. Agricultural and Food Policy. 4th ed. Prentice Hall, Upper Saddle River, NJ.

Madden, J.P., and T.L. Dobbs. 1990. The role of economics in achieving low-input farming systems. p. 459–477. *In* C.A. Edwards et al. (ed.) Sustainable agricultural systems. Soil and Water Conservation Society, Ankeny, IA.

Mullarkey, D., J. Cooper, and D. Skully. 2001. "Multifunctionality" and agriculture: Do mixed goals distort trade? Choices 16:31–34.

Oglethorpe, D.R. 1995. Sensitivity of farm plans under risk-averse behavior: A note on the environmental implications. J. Agric. Econ. 42:227–232.

Prato, T. 1998. Natural resource and environmental economics. Iowa State Univ. Press, Ames.

Pretty, J.N., C. Brett, D. Gee, R.E. Hine, C.F. Mason, J.I.L. Morison, H. Raven, M.D. Rayment, and G. van der Bijl. 2000. An assessment of the external costs of UK agriculture. Agric. Syst. 65:113–136.

Pretty, J., C. Brett, D. Gee, R. Hine, C. Mason, J. Morison, M. Rayment, G. van der Bijl, and T. Dobbs. 2001. Policy challenges and priorities for internalizing the externalities of modern agriculture. J. Environ. Plann. Manage. 44:263–283.

Smolik, J.D., T.L. Dobbs, and D.H. Rickerl. 1995. The relative sustainability of alternative, conventional, and reduced-till farming systems. Am. J. Altern. Agric. 10:25–35.

Stewart, L., N. Hanley, and I. Simpson. 1997. Economic valuation of the agri-environmental schemes in the United Kingdom. Report to HM Treasury and the Ministry of Agriculture, Fisheries and Food. Environmental Economics Research Group, University of Sterling.

Whitby, M. 2000. Challenges and options for the agri-environment: Presidential address. J. Agric. Econ. 51:317–332.

Zilberman, D., and M. Marra. 1993. Agricultural externalities. p. 221–267. *In* G.A. Carlson et al. (ed.) Agricultural and environmental resource economics. Oxford University Press, New York.

7 Community Dynamics and Social Capital

CORNELIA BUTLER FLORA

North Central Regional Center for Rural Development
Iowa State University
Ames, Iowa

Communities are critical for successful agroecosystems. In addition to increasing private profit, they provide contexts that support or discourage sustainability and processes that augment participation in the development of public goods. Further, communities are impacted by surrounding agroecosystems through the production of goods and services, as well as any changes in their production. These impacts, both physical and social, are highly interactive. It is all too easy to view the social impact as a residual category, both in terms of causes and effects of agroecosystems; however, social variables can be measured on different scales as vital components at different levels of agroecosystems. In this chapter, we will deal with community-level measures. An important part of community, which influences the context, processes, and impacts, is social capital. Social capital can be an individual, community, state, or even national characteristic. In order to understand agroecosystems and measure social impacts, social capital is usefully conceptualized at the community level.

Behaviors and decisions that impact agroecosystems are embedded in communities. A general understanding of the intersection between individual and firm decision making becomes valuable as we identify the social system that impinges upon and is conversely affected by biological and physical systems.

WHY DO PEOPLE DO WHAT THEY DO?

For communities and agroecosystems, it helps to frame a social model of behavior in terms of social control. Visually, social control is a pyramid (Fig. 7–1), with a base comprising recognizable democratic formula, where social control simply conforms to norms. At the top of the pyramid is the use of coercion and force, which serves best as a threat and is rarely implemented; this formula corresponds to an old military adage related to using the least possible force in each battle in order to win the war.

On one side of the pyramid are positive sanctions, with negative ones on the other. Society works best when positive sanctions are used. Yet without the possibility of negative sanctions, positive sanctions are not as effective in enhancing society-approved behavior.

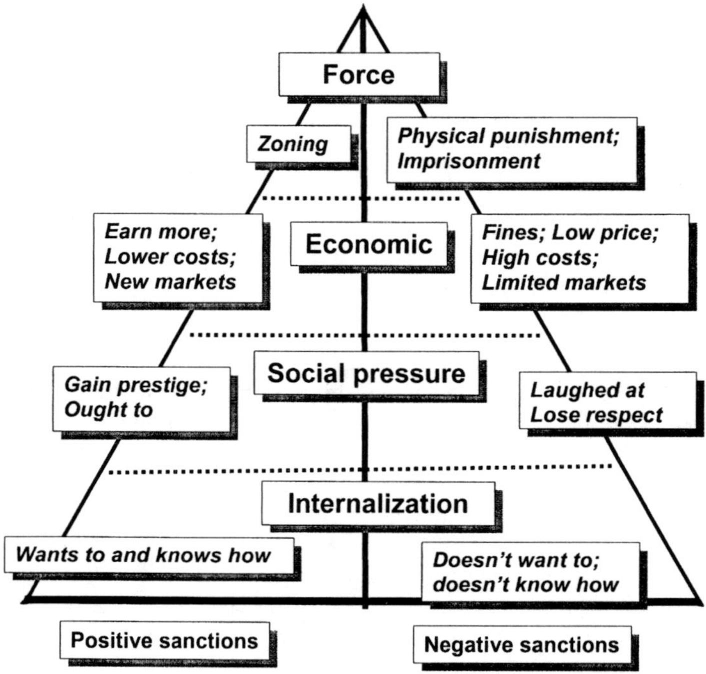

Fig. 7–1. The social control pyramid.

At this point, it is important to note that sustainability of ecosystems is not necessarily defined as a societal good. In the past, nature has been viewed as an entity to be conquered, not worked with (Gasteyer and Flora, 2000a; Merchant, 1980; Vileisis, 1997). Thus, when we look at social control and human behavior, we have to take into account that social control mechanisms can either enhance the sustainability of or support the deterioration of agroecosystems.

The best mechanisms of social control are those that are practically invisible. Individuals are socialized to do the right thing. Societal norms regarding interaction with the land are *internalized.* The culture of the firm takes pride in being "green" and delivering on the double bottom line (environment and economy). Salamon and her colleagues demonstrated how the attitudes of the family of origin—winning the prize for conservation vs. obtaining the prize for the highest yields along with the pride of getting the biggest machinery first—influence how farm managers design and manage agroecosystems (Salamon et al., 1998). Their careful ethnographic research has documented how community, which in many farming areas is coterminous with ethnicity, determines the purpose of farming. As an example, Salamon (1992) contrasted communities dominated by "Yankee Farmers" compared with those of "Yeoman Farmers."

In dealing with different types of agroecosystems, the assumption of USDA-based programs is that people will "do the right thing" to meet society's expectations, meaning either conservation or maximum production, if they know how. Teaching them how to do the right thing is the sole purpose of these programs. Of course, when values are not shared, when the individual or firm does not want to conform, all the education in the world will not bring about change in management of the agroecosystem.

When socialization disconnects from current societal norms or incomplete internalization has occurred (e.g., the individual or firm is drawn to deviate from the norm because of short-term profits or a vision of an alternative future), the next level of social control is activated: peer pressure. Because humans judge their self-worth by the worth attributed to them by significant others, this can be very powerful. Initially, many farmers were hesitant to pursue no-till farming, fearing they may be labeled "bad farmers." Adoption of Integrated Pest Management (IPM), which is based on threshold densities of pests, has been hampered by traditional zero tolerance for weeds in many rural communities. Accordingly, gaining respect or avoiding ridicule helps convince people and firms to execute tasks that might not be internalized.

In small, static communities, where everyone knows everyone else and where the land operator is the landowner, norms that define a "good" agroecosystem are relatively easy to enforce. What one does is equal to what one is, and both are extremely visible—unlike the office worker who is known more by what that worker consumes (Barlett, 1993). But when ownership is separated from management, and when what serves the public good is not internalized and is viewed by the dominant culture as in conflict with private profit, the next level of social control comes in: economic sanctions. For changes in agroecosystems to occur, groups that present alternative reflections of the individual or firm are critical (Meares, 1997; Hassanein, 1999).

Economic sanctions around agroecosystems are quite contradictory in the 2002 Farm Security and Rural Investment Act (Farm Bill) (USDA, 2002). On the one hand, maximum production of a few crops is highly rewarded economically as federal programs reduce both natural and economic risks through emergency programs and a variety of mechanisms that alter loan prices and loan deficiency payments. On the other, there are economic rewards in terms of cost sharing and direct payments for farmers who are good stewards and provide ecosystem services. But because positive sanctions that promote production encourage the very practices that the conservation payments try to discourage, the result is generally one of little net gain. In fact, the dollars invested in support payments far exceed those dedicated to conservation measures. However, communities around agroecosystems can provide positive sanctions, such as property tax breaks or direct payments, to encourage farmers to produce public goods through changing land use. Accordingly, communities of interest are willing to pay higher prices for crops produced in agroecosystems viewed as healthier for the planet.

Negative economic sanctions also are invoked in agroecosystems. Confined Animal Feeding Operations (CAFOs) that have manure spills can be fined. Landowners and managers whose ditches are filled by erosion can be charged for cleaning them out. The USEPA, accustomed to working with firms where management and ownership are separated, has had great success with industry by using negative economic sanctions. Indeed, after protesting loudly that requirements were far too costly to possibly implement, many firms beat the time line established by the federal government to clean up air and water, discovering that pollution is waste that costs money and cuts into profits, which aggravates stockholders (Schultze, 1999; Richards, 1997; Richards and Pearson, 1998). Thus negative economic sanctions, the threat of serious fines, combined with outcome-based (as opposed to design-based) standards led to creativity and innovation by industry to protect ecosystem health (Carpentier and Bosch, 1999). Sometimes the market itself imposes negative economic sanctions on ecosystem management practices seen as harmful to the agroecosystem. For example, McDonald's has implemented strict standards for raising chickens, causing poultry growers, as part of a more intensive agroecosystem, to lose their market unless their practices changed.

Finally, there are the sanctions of force. Planning and zoning are examples of such sanctions that influence agroecosystems. Certain activities can be done in certain places and are forbidden in others. In many states, there is serious debate around the traditional exemption of agriculture from planning and zoning, especially as agricultural production practices become more concentrated and more industrial (Tweeten and Flora, 2001; Committee to Review the Role of Publicly Funded Research on the Structure of Agriculture, 2002). Some states and counties have imposed moratoria on the construction of CAFOs, an example of using the strongest sanction possible in the case of agroecosystems.

By understanding the norms and the sanctions that influence different levels of agroecosystems, we can predict the level at which change will have to be made in mechanisms of social control, voluntary (internalization, peer pressure, and positive economic sanctions) or involuntary (negative economic sanctions and any force).

Since the norms, their enforcement, and willingness to change are critical in influencing the forms that agroecosystems will take, it is important to assess the degree to which communities at various levels are open to consideration of alternative agroecosystems, new ideas, and systems interactions. Social capital is a useful tool for such analysis.

SOCIAL CAPITAL AND CONTEXT

Social capital involves relationships among individuals and groups. Those relationships can vary as to the degrees of mutual trust, reciprocity, shared norms and values, common goals, and sense of a shared future. Moreover, those relationships can be very dense and internally oriented within the community (or segments of the community), which we refer to as *bonding social capital*. Conversely, they can reach out beyond the "comfort zone" to different kinds of people and groups in different places, which is referred to as *bridging social capital.* Communities can vary in both (see Fig. 7–2 and 7–3). When both are low, people are atomized, with little interest in the public good. When bridging capital is low and bonding capital is high, there is often great internal factionalism and resistance to both compromise and change. When bridging capital is high, but bonding capital is low, a small group often controls the community—their particular outside connections allow them to reward loyal factions of the community. When both bridging and bonding capital are high, there are high possibilities of internal collaboration, participation and open-

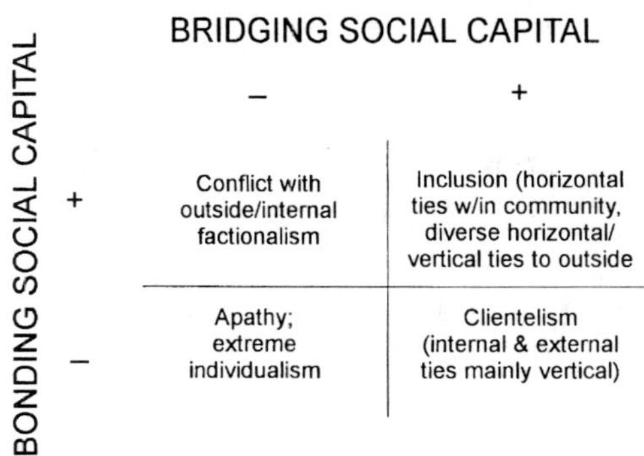

Fig. 7–2. Community social capital typology. Adapted from Flora and Flora, 2002.

ness to constantly improve the collective condition. Flora and Flora (1993) described this as *"entrepreneurial social infrastructure"* (ESI).

Entrepreneurial social infrastructure, as the community context for sustainable agroecosystems, is based on three major components:

- The legitimacy of alternatives
- Internal and external networks
- Ability to mobilize resource

When ESI is high, the community and its members are able to separate means from ends. For example, instead of saying, "we need a new water treatment facility," the discussion starts with: "We have water quality problems in terms of unhealthy levels of nitrate. What are different ways we can reduce nitrate levels?" Framing a question around the main issue, rather than presenting a ready-made "solution," allows for consideration of ecosystem services and the cost-effectiveness of building a nitrate removal plant versus rewarding land managers and land owners who reduce the use of N fertilizer and introduce barriers to the flow of nitrate to surface water in order to change their agroecosystem.

In order to get multiple perspectives on issues, it is important to have communications and linkages both inside and outside the group and community. When

Fig. 7–3. Community social capital typology and change. Adapted from Flora and Flora, 2002.

sources of knowledge on sustainable agroecosystems are not available from traditional knowledge sources, links to alternative sources are particularly important. Kroma and Flora (2001) revealed that when extension and input dealers did not have the information, new networks were needed, both formal and informal. These networks are two-way, as knowledge is both gained and shared. Further, forming new value chains for marketing both new agricultural products (instead of just commodities) and for marketing agroecosystem services is a critical role that will lead to the greater connectivity required for more sustainable agroecosystems.

Ability to mobilize resources is critical for agroecosystem enhancement. Change can occur from asset deterioration or from investment in the asset to create new resources. The first land managers who sought innovation for more sustainable agroecosystems found it difficult to attain credit and technical assistance to make the "risky" shift. Off-farm people did not understand the new systems and therefore did not support them. New alliances that supported alternative resource flows, including establishing machinery sharing arrangements and seed exchanges for heirloom varieties, are ways that agroecosystems are enhanced. The ability to mobilize nonfarm interest in recognizing and rewarding public goods and service production has been a powerful force for conservation since the turn of the 20th century.

A national study of rural communities found ESI predicted community self-development, that is, community members and firms working together in public–private collaborations for economic development through local initiatives (Flora et al., 1997). Local innovation in agriculture was related to level of ESI in my study of the Great Plains and the Corn Belt (Flora, 1995). High ESI does not necessarily lead to more sustainable agroecosystems. Indeed, when the connections are with high input networks, the results will be the opposite. Farmers will ignore the indicators of ecosystem deterioration, such as the fact that no one drinks the water from the local water supply heavily laced with atrazine (6-chloro-N^2-ethyl-N^4-isopropyl-1,3,5-triazine-2,4-diamine) because the entire community views farming row crops to the waters edge as an acceptable practice (strong internal networks), and cultural practices are supported by the commodity buyers and the input suppliers (single, strong external networks).

PROCESSES TO BUILD SUSTAINABLE AGROECOSYSTEMS

Engaging in behavior to change a local situation, whether by an individual or organizations, requires *discovering* the following:

- The current situation is problematic.
- The current situation is related to human activity.
- Alternatives to current behavior and its consequences are possible.

Once that has happened, individual, community, and institutional actors can engage in mobilizing resources to bring about change. Awareness of a threatening situation at an international, national, or even state level may not lead to awareness of local issues and alternatives nor lead to local action. Discovery, learning, and engagement are intrinsically social activities. However, if discovery, learning, and engagement are not widely occurring within a local area, action related to improving ecosystem health, while at the same time dealing with issues of social equity and economic vitality, is unlikely to be sustainable.

If social capital can serve to either enhance or detract from agroecosystem health, what can be done to promote social capital to enhance the agroecosystem? We have conducted a meta-analysis of participatory natural resource management strategies proposed by both practitioners (in their manuals) and by theorists (in their scholarly publications). From this, we derived eight basic processes that both build ESI and healthy agroecosystems; we consider each of these in the following sections.

Context Specificity

Contemporary, industrial agroecosystem management is often based on technologies and methods that can be applied in many places simultaneously. For example, specific agricultural technologies, such as hybrid seeds or conservation tillage, were widely distributed to farmers as improvements with relatively little adaptation to the local ecological, economic, and social context. While these technologies improved the quality of life for some farmers, they also contributed to ecological and social problems by encouraging monoculture (Hazel and Ramsamy, 1991; Pretty, 1995). Best management practices (BMPs) are assumed to be goods in themselves, rather than useful in context. Some programs measure success by the presence of BMPs, rather than by the actual ecological conditions achieved. This mechanistic world view has perpetuated industrial agriculture (Keller and Brummer, 2002), where the local is overcome and dominated, rather than promoted and adapted to.

Participatory approaches often emerge out of an effort to create development processes deeply rooted in the local context and to recognize the uniqueness of place. Communities carry out activities, such as transect walks, where community members traverse the various ecosystems and microclimates of their community or farm to understand the ecological–environmental, social, and economic aspects of their community. Together, the community and outside experts identify issues, propose solutions, and evaluate actions on the basis of an understanding of the local context. Outside technology may be brought in, but is adopted based on, rather than in spite of, the local context (Savory, 1999; Pretty and Chambers, 1994).

Involvement of Diverse Perspectives

Agroecosystem management tends to address problems from a narrow range of options. It has been assumed that the only people to be involved are land managers. For example, in designing irrigation systems, only those who practiced irrigated agriculture were included, but water affects everyone. In Victoria, Australia, when other stakeholders in the watershed were consulted, including mothers' groups, housing organizations, city officials, and recreation and sports clubs, the way the irrigation system was designed changed radically, because shifting water use affected everything.

Often, citizens in farming communities believe they must choose between preserving the environment and economic well-being. For example, one either has large industrialized hog confinement operations, or the whole industry moves elsewhere. The debates engage the polar voices, without consideration of the alternatives between all or nothing. Alternatives to such zero sum thinking (if you win, I lose) are not developed, and other voices that may present less stark alternatives are excluded from the discussion. By seeking perspectives of unconventional participants in decision making, the community may develop other options that allow solutions that serve multiple interests, rather than trading off one for the other (see Freire, 1970; Chambers, 1983; Habermas, 1979). For example, by including elected city officials, county planners, and sports groups in discussions about agroecosystems in the context of the county or watershed systems, new perspectives can be gained.

Including diverse perspectives is more than having at the table several different folks who each make her/his own case. It requires an open process whereby community members meet and identify major concerns and existing community assets. They then identify possible solutions for those concerns, using local assistance as much as possible. Numerous methodologies, such as cognitive maps, have been developed to facilitate group inquiry into problems and possible solutions (see Slim and Thompson, 1995; Rocheleau, 1994).

Collective Vision and Focus on Outcomes

When place is central to the discussion and diverse perspectives are presented that address the conditions of place, it becomes possible to develop a shared vision of what a better place would be—the desired future conditions (Lightfoot and Okalebo, 2001).

A necessary component of empowering communities to approach development and management as a long-term learning process involves encouraging communities to imagine desired outcomes, rather than outputs. Conventional development has been based on the construction of outputs, the physical results of activities, infrastructure, and events. For instance, the attempts to alleviate poverty in the 1960s often involved the construction of infrastructure, such as roads and sewer systems. While these outputs produced countable products for the dollars spent, they often did not affect the social structures in persistently poor communities that kept poor people marginalized, disenfranchised, and impoverished. By deciding on desirable outcomes as a first stage of the project, activities can be designed, evaluated, and amended according to those outcomes, furthering the achievement of project goals (Engel, 1997; Flora 1998).

Monitoring

Both practitioners and academics are developing monitoring systems that are applicable at the community level. Monitoring depends on community agreement on social, economic, and environmental goals, as well as on desirable outcomes of activities. These negotiated outcomes are generally built from a community visioning process, often present in strategic plans. On the basis of these, the community then participates in a process of developing indicators and a framework for monitoring the community's activities, assessing whether the activities and their outputs lead toward the desired outcomes.

Whether the monitoring is performed using locally developed indicators or those available through academic or agency databases depends on the availability of appropriate information, how that information is going to be distributed, and the purpose of monitoring. For some communities, local indicators, such as a wade-in sampling systematically done on an annual basis to check with turbidity, are more effective in spurring community action than scientific indicators (see Gasteyer and Flora, 2000a). Other contexts (such as large urban areas) require the legitimacy of monitoring using scientifically accepted indicators and experts implementing them (Hart, 1999; Innes and Booher, 1999; Innes, 1996; Flora, 1998; Andrews, 1996).

Systematic Learning through Negotiating Evidence

Moving toward the future vision of the agroecosystem requires more than a collective definition. It is critical that there is constant learning and adaptation to new insights and constantly changing conditions by all stakeholders. Communities develop a process for understanding the local economic, social, and natural system and analyzing how actions and policies impact that system. By establishing processes that encourage constant learning and adaptation, communities debunk the myth of the silver bullet that solves problems in perpetuity. For example, "Guarantee us a price for our single crop and get the government out of our face," is a silver bullet proposed by some farmers regarding management of their monocropped agroecosystem. In contrast, communities engaged in continuous learning are more apt to actively address important issues, as well as to develop a systematic approach to measure the impacts of those actions. Community members are prone to value their own insights and knowledge systems if they are assisted in developing an efficient way of learning about the ecological, social, and economic system in which they live—and evaluating actions in reference to those systems. Ultimately agroecosystem management should be about empowering communities to improve their quality of life through better investments in their natural resources. Through developing a systematic learning approach, community members potentially develop the tools for ongoing analysis of the management of natural resources, rather than having to depend on outside experts (Pretty and Chambers, 1994; Hincliff et al., 1999; Innis and Boher, 1999).

Conventional development involves decisions made by a small group within the community, usually in collaboration with outside experts (sometimes scientists) who interact with other development experts or researchers, but not with others in the local community. Many participatory approaches, in contrast, explicitly attempt to widen the circle of decision making to involve more of the community. Group inquiry involves an open forum where community members meet and identify the major issues of concern and existing community assets. Concerns and assets serve as the base for a process of identifying possible solutions for those concerns, using local tools as much as possible.

Many of the approaches to participatory development emphasize that community development and management of natural resources involve a commitment to long-term management, rather then quick fixes and immediate technical solutions to existing problems. It has been necessary, then, to design approaches to participatory development, planning, and natural resources management that are based on long-term, sustained learning and action. Often, this involves activities that reward accomplishments by members of the local community. For instance, the project might involve organized field visits to local farms where the owners are trying innovative approaches. Other initiatives might also develop, and an action step is to publish the indicator frameworks developed by the community as a way to monitor progress toward shared goals. Community learning sessions are a constructive way for community members to learn from each other about the history, ecology, and society in which they live. In other cases, experts train residents of the community in research techniques so they can carry out future research of interest. This is all done in order to empower the community in long-term development decision making for ecosystem management (Innes, 1996, 1998; Guite and Thompson, 1999; Engel, 1997; Rocheleau, 1994).

Neutral Facilitators

While empowerment of local citizens is critical for sustainable agroecosystems, many approaches also recognize that external facilitators (either from nonlocal government agencies or nonlocal nongovernmental organizations) can play a key role in shaping agroecosystems at the community level. These neutral agents provide not only technical and scientific knowledge, but also an essential outsider's view that will illuminate certain factors or patterns in the community. They can provide alternatives to the local assumptions about how to manage natural resources and economic, social system options. They may also have the freedom to challenge existing social hierarchies, taboos, and power structures—allowing for future local processes that are more inclusive and participatory. Some of the literature on participatory approaches has explicitly tried to outline the appropriate role for external agents in community development and the points in the process where they should be more or less dominant (Rocheleau, 1994; Chambers, 1983; Engel, 1997).

Participatory Contract

Developing a participatory contract for a project ensures that the rights and responsibilities of community members, researchers, and outside managers are transparent and explicit. All members of the initiative sign an agreement that clearly states their action plan and expectations for reimbursement (monetary or in kind) for their efforts. The contract also specifies when each party is empowered to terminate the agreement. In this way expectations of what the process will produce are clear from the beginning of the initiative (Pretty, 1994).

Evaluation in the Context of Community

Often participatory approaches involve evaluation both by the end users and the researchers or technical managers of the development or management initiative. This allows community members to voice opinions about the initiative, whether it accomplished anything important for the community, and what would make the process more useful in the future. Comparing the community and outside agents' perspectives on the project may also illuminate differences in perception and lead to improved interactions between these groups in the future (Guite and Thompson, 1999; Slim and Thompson, 1995).

IMPACTS OF AGROECOSYSTEMS ON COMMUNITIES

Just as the influences that impact agroecosystems exceed the individual experiences of land managers or the firms that hire them (Smith, 1998), those impacts are relatable to four capitals in a community, watershed, or region: human, social, natural, and financial–built capital (Flora, 2001).

The impacts of agroecosystems are measurable, and like the processes involved, measurement is best accomplished through community participation in order to ensure that the measures are meaningful to the stakeholders whose decisions, directly and indirectly, impact land use. There are many ecosystem impact measurement tools (Flora et al., 1999; Hart, 1999; North Central Regional Center for Rural Development, 1999; Toupal and Johnson, 1998).

We have found that for research and for agroecosystem development, using a menu for precise measures with general indicators around the four capitals is useful and parsimonious. The secret of using indicators is to remember that each indicator represents only a piece of the desired outcome. Such indicators must be (i) meaningful, (ii) linked to human action, (iii) relatively easily measured, and (iv) not chosen on the basis of the ease of their measurement.

We found that the following basic sets of indicators are inclusive enough to give a holistic picture of impacts and outcomes.

Increased use of the knowledge, skills, and abilities of local people (human capital)

Does the agroecosystem encourage the identification of the skills, knowledge, and abilities of local people? Does it encourage an increase in the skills, knowledge, and ability of local people? Does it facilitate recombining the skills, knowledge, and abilities of local people? Or, does the agroecosystem in essence de-skill local people, depending on experts and outside interventions to solve problems or even make it function?

Strengthened relationships and communication (social capital)

An agroecosystem can increase the flow of information within the community, watershed, or region, or it can decrease it, as knowledge and technologies are privatized and individual solutions are sought. New networks within the community can form in response to the opportunities of an agroecosystem's multiple functions, or they can deteriorate as the single production function is stressed. Agroecosystems can stimulate new connections outside the community with civil society groups, including environmental and recreational organizations, and market groups, including specialized distributors. They can encourage single linkages to Congress to ensure subsidies for the few crops that have traditionally been grown in the area.

Increased flexibility, innovation, and adaptation (social capital)

Agroecosystems can encourage innovation as they constantly respond to changing conditions, or they can increase rigidity as farmers seek to replicate conditions of an idealized agrarian past. Diverse agroecosystems encourage flexibility in response to market and environmental changes, whereas large monocultural systems resist change at all levels. Agroecosystems that are static can encourage a victim mentality—everyone (particularly the government and the liberal press) is out to get us and always blames us for everything—or a cargo cult mentality—waiting for someone to build a factory or guarantee a price and thus solve the local problems.

Sustainable, healthy ecosystem with multiple community benefits (natural capital)

Agroecosystems can be a catalyst for communities planning and acting in concert with the environment, or they can be the reason to "tame" nature. Agroecosystems can provide a place for groups with different land use objectives to find common ground, or they can be a source of constant conflict for the community. Additionally, agroecosystems can be used for multiple community benefits, or they enrich a few firms or farmers only.

Appropriate diverse and healthy economies (financial and built capital)

Agroecosystems can increase community poverty, as for example monoculture cotton (*Gossypium hirsutum* L.) and soybean [*Glycine max* (L.) Merr.] has in the rich Mississippi Delta. Conversely, they can provide opportunities for poverty reduction, both through providing alternative income sources and healthy food for self-provisioning. Agroecosystems can contribute to local business diversity if they are distinct and small enough to be locally served. Alternatively, they can contribute to the decline of both agricultural and nonagricultural businesses in an area by using few locally purchased inputs and selling outside the area. Agroecosystems can contribute to business efficiency by encouraging linkages to better market signals and more efficient ways of utilizing labor and capital, or they can be based on low wage labor and only respond to government price supports. Agroecosystems can increase local residents' wealth by retaining value in the area, or they can export the wealth through high volume, low value crops.

CONCLUSION

Social factors greatly influence the context under which agroecosystems evolve and are maintained, as well as the processes by which they remain static or are constantly changing. Agroecosystems, in turn, have impacts far beyond agricultural production and even the provision of ecosystem services. The ways they are organized within the context at the local and landscape levels impact how communities respond to constant change, assess opportunities, build on local assets, and work together.

Building bridging and bonding social capital between land managers and the communities they surround can enhance sustainability. But, unless ecosystem health is considered explicitly as an outcome, both bridging and bonding social capital can further separate humans from their physical and biological environment. Solutions that are technically sustainable, but do not engage the broader community, are much less likely to be widely implemented.

STUDY QUESTIONS

1. Using Fig. 7–1 and its description at the beginning of this chapter, relay an experience that corresponds with the pyramid on some level. What was the outcome of that experience?

2. What are some possible positive impacts of social capital in a community? Negative?

3. Consider the impacts of sanctions, as well as social control mechanisms and internalized social behavior. Invent a fictitious community or use one you have been a part of and explain how these forces operate in the community. The community needs to stop farming crops so close to the water edge. Outline possible personalities or groups that are part of the community, for example, how do some of the older residents ideas mesh with younger residents' outlook for the future? What could be possible outcomes of having a discussion with different members of the community? Who has the control, and how is it disseminated? What are the desired outcomes? Is a neutral agent necessary?

4. Entrepreneurial social infrastructure does not always lead to sustainable agroecosystems; this is illustrated with an example in the Social Capital and Context section of this chapter. How can ecosystem deterioration be stopped or turned around? How can people be motivated to "do the right thing" without someone intruding on their "cultural practices?"

5. How could a discussion between an athletic coach and county planners be facilitated? What multiple perspectives might be addressed?

6. How might neutral agents be received in a strong community with a lot of social capital? It is possible that the agents' alternatives to local assumptions could face defensive reactions. How does one recognize an appropriate time to invite a neutral agent to the table? How does one troubleshoot in these situations, if necessary?

7. What elements need to be in place for an agroecosystem to become sustainable?

REFERENCES

Andrews, J. 1996. Going by the numbers [online]. Available at www.planning.org/planningpractice/1996/Sept96.html (accessed 16 July 2003, verified 1 Oct. 2003). American Planning Association, Washington, DC.

Barlett, P.F. 1993. American dreams, rural realities: Family farms in crisis. Univ. of North Carolina Press, Chapel Hill.

Carpentier, C.L., and D.J. Bosch.1999. Design versus performance standards to reduce nitrogen runoff: Chesapeake bay watershed dairy farms. *In* F. Casey et al. (ed.) Flexible Incentives for the adoption of environmental technologies in agriculture. Kluwer Publishing Company, Boston, MA.

Chambers, R. 1983. Rural development. Putting the last first. Longman Publishing Company, London.

Committee to Review the Role of Publicly Funded Research on the Structure of Agriculture. 2002. Publicly funded research and the changing structure of U.S. agriculture. National Academy Press, Washington, DC.

Engel, P. 1997. The social organization of innovation: A focus on stakeholder interaction. Royal Tropical Institute, Amsterdam.

Flora, C., and J.L. Flora. 1993. Entrepreneurial social infrastructure: A necessary ingredient. Ann. Acad. Soc. Polit. Sci. 529:48–58.

Flora, C., and Flora, J. 2002. Community dynamics and types of capital. *In* D.L. Brown and L. Swanson. (ed.) Challenges for rural America in the 21st century. Penn State Univ. Press, University Park.

Flora, C.B. 1995. Social capital and sustainability: Agriculture and communities in the great plains and the corn belt. Res. Rural Sociol. Dev. 6:227–246.

Flora, C.B. 1998. Community building for a healthy ecosystem. [Online.] Available at http://www.ag.iastate.edu/centers/rdev/newsletter/fall98/contents.html (verified 12 July 2003). Rural Dev. News 22:1–2.

Flora, C.B. 2001. Shifting agroecosystems and communities. p. 5–14. *In* C. Flora. (ed.) Interactions between agroecosystems and rural communities. CRC Press, Boca Raton, FL.

Flora, C.B., M. Kinsley, V. Luther, M. Wall, S. Odell, S. Ratner, and J. Topolsky. 1999. Measuring community success and sustainability. RRD 180. [Online.] Available at http://www.ncrcrd.iastate.edu/Community_Success/about.html (verified 12 July 2003). North Central Regional Center for Rural Development, Ames, IA.

Flora, J.L., J.S. Sharp, C.B. Flora, and B. Newlon. 1997. Entrepreneurial social infrastructure and locally-initiated economic development. Sociol. Q. 38:623–645.

Freire, P.1970. Cultural action for freedom. Monogr.1. Harvard Educational Review Center for the Study of Development and Social Change, Cambridge, MA.

Gasteyer, S., and C.B. Flora. 2000a. Measuring PPM with tennis shoes: Science and locally meaningful indicators of environmental quality. Soc. Nat. Res. 13:589–597.

Gasteyer, S., and C.B. Flora. 2000b. Modernizing the savage: Colonization and perceptions of landscape and lifescape. Sociol. Ruralis 40:128–149.

Guite, I., and J. Thompson. 1999. Sustainability indicators for analysing the impacts of participatory watershed management programmes. p. 13–26. *In* F. Hincliff et al. (ed.) Fertile ground: The impacts of participatory watershed management. IT Publ., London.

Habermas, J. 1979. Communication and the evolution of society. T. McCarthy, translator. Beacon Press, Boston, MA.

Hart, M. 1999. Guide to sustainable community indicators. Hart Environmental Data, North Andover, MA.

Hassanein, N. 1999. Changing the way America farms: Knowledge and community in the sustainable agriculture movement. Univ. of Nebraska Press, Lincoln.

Hazel, P.B.R., and C. Ramsamy. 1991. The Green Revolution reconsidered: The impact of high-yielding rice varieties in South India. Johns Hopkins Univ. Press, Baltimore.

Hincliff, F, J. Thompson, J. Pretty, I. Guijit, and P. Shah (ed.) 1999. Fertile ground: The impacts of participatory watershed management. IT Publications, London.

Innes, J.E. 1996. Planning through consensus building: A new view of the comprehensive planning ideal. J. Am. Plann. Assoc. 62:460–481.

Innes, J.E. 1998. Information in communicative planning. J. Am. Plann. Assoc. 64:52–63.

Innes, J.E., and D.E. Booher. 1999. Indicators for sustainable communities: A strategy building on complexity theory and distributed intelligence. Institute of Urban and Regional Development, Univ. California, Berkeley.

Keller, D.R., and C. Brummer. 2002. Putting food production in context: Toward a postmechanical agricultural ethic. BioScience 52:264–271.

Kroma, M., and C.B. Flora. 2001. An assessment of SARE-funded farmer research on sustainable agriculture in the North Central U.S. Am. J. Altern. Agric. 16(2):73–80.

Lightfoot, C., and S. Okalebo. 2001. Vision based action planning report on a learning process for developing guidelines on vision based sub-county environmental action planning. International Support Group, Montpelier, France.

Meares, A.C. 1997. Making the transition from conventional to sustainable agriculture: Gender, social movement participation, and quality of life on the family farm. Rural Sociol. 62:21-47.

Merchant, C. 1980. The death of nature: Women, ecology, and the scientific revolution. Harper and Row, San Francisco.

North Central Regional Center for Rural Development. 1999. Social indicators: An annotated bibliography on trends, sources and development, 1960–1998. [Online.] Available at http://www.ncrcrd.iastate.edu (verified 12 July 2003). NCRCRD, Ames, IA.

Pretty, J. 1994. Regenerating agriculture: Policies and practice for sustainable growth and self reliance. John Henry Press, Washington, DC.

Pretty, J. 1995. Regenerating agriculture: Policies and practice for sustainable growth and self reliance. John Henry Press, Washington, DC.

Pretty, J., and R. Chambers. 1994. Towards a learning paradigm: New professionalism and institutions for sustainable agriculture. p.182–202. *In* I. Scoones and J. Thompson (ed.) Beyond farmer first. Intermediate Technology Publications, London.

Richards, D.J. 1997. The industrial green game: Implications for environmental design and management. National Academy of Engineering, National Academy Press, Washington, DC.

Richards, D.J., and G. Pearson. 1998. The ecology of industry: Sectors and linkages. National Academy of Engineering, National Academy Press, Washington, DC.

Rochleau, D. 1994. Participatory research and the race to save the planet: Questions, critiques and lessons from the field. Agric. Human Values 2(2,3):4–25.

Salamon, S. 1992. Prairie patrimony: Family, farming, and community in the Midwest. Univ. of North Carolina Press, Chapel Hill.

Salamon, S., R.L. Farnsworth, and J.A. Rendziak.1998. Is locally led conservation planning working? Rural Sociol. 63(2):214–234.

Savory, A. 1999. Holistic management: A new framework for decision making. Island Press, Washington, DC.

Schultze, P.C. (ed.) 1999. Measures of environmental performance and ecosystem condition. National Academy of Engineering, National Academy Press, Washington, DC.

Slim, H., and P. Thompson.1995. Listening for a change: Oral testimony and community development. New Society Publishers, Philadelphia, PA.

Smith, G. 1998. Are we leaving the community out of sustainable community development: An examination of approaches to development and implementation of indicators of rural community sustainability and related public participation. Int. J. Sustainable Dev. World Ecol. 5:82–98.

Toupal, R.S., and M.D. Johnson. 1998. Conservation partnerships: Indicators of success. Technical Report. Social Science Institute, USDA-NRCS, Madison, WI.

Tweeten, L.G., and C.B. Flora. 2001. Vertical coordination of agriculture in farming-dependent areas. Task Force Rep. 137. Council for Agricultural Sciences and Technology, Ames, IA.

USDA. 2002. Farm security and rural reinvestment act of 2002. Available at www.usda.gov/farmbill/index.html (accessed 17 July 2003, verified 1 Oct. 2003). USDA, Washington, DC.

Vileisis, A. 1997. Discovering the unknown landscape: A history of America's wetlands. Island Press, Washington, DC.

8 Conservation within Agricultural Landscapes

THOMAS E. SCHUMACHER AND DIANE RICKERL

South Dakota State University
Brookings, South Dakota

One of the primary goals of conservation in agriculture is the use of management practices that will sustain agriculture as a basic component of the rural landscape (i.e., sustainable agricultural landscapes). A desirable goal of conservation is to use farming practices that will sustain ecological goods and services of the rural landscape. Ecological goods and services include clean drinking water, wildlife habitat, human habitat, clean air, aesthetics, pollution-free streams, and healthy lakes. The discipline of agroecology attempts to meet these goals by applying ecological concepts and principles to the design and management of sustainable agricultural landscapes (Gliessman, 1998).

For our purposes, a landscape is an area on the scale of several kilometers composed of interacting ecosystems (also referred to as landscape elements) that are repeated because of geology, landform, soils, climate, biota, and human influences throughout the area (Gregorich et al., 2001). There are three types of landscape elements within a landscape: patches, corridors, and matrix (Forman and Godron, 1986). An agricultural landscape is a landscape in which the dominant landscape element (also known as the matrix) is agricultural production. A patch is an element of the landscape that is not dominant but occurs repeatedly within the matrix. There are likely to be several kinds of patches within an agricultural landscape, including wetlands, woodlots, lakes, farmsteads, wildlife refuges, community meeting houses, cemeteries, grain storage facilities, rural business sites, and towns. Corridors are elongated connections between patches and matrix components (e.g., fields). Examples of corridors that are likely to be found in agricultural landscapes include roads, windbreaks, fence lines, riparian buffers, animal trails, streams, ditches, power lines, right of ways, and grassed waterways.

In this chapter we introduce a new term called *repetitive landscape unit* (RLU), which is the minimum land area that contains landscape elements in the same proportion as the larger landscape. A landscape is composed of many repetitive landscape units. In agriculture an analogous term is the soil management unit.

Landscape structure is determined by the distribution of energy, materials, and organisms within and among these landscape elements and RLUs. This distribution is in turn determined by the sizes, shapes, numbers, kinds, and configurations of landscape elements that make up the RLUs (Forman and Godron, 1986). More detailed descriptions of landscape structure and its relationship to natural and social resources are given in Chapter 10 (Francis et al., 2004).

Agriculture by definition is dependent on human involvement and activity. The development of agricultural landscapes started 10 000 yr ago in the Pre-Pottery Neolithic Near-East with the successful domestication of major grain and legume crops (Lev-Yadun, 2000). Agricultural sustainability has varied geographically and through time. There are instances where agriculture has vanished from the landscape (Lowdermilk, 1953). There are also examples where agricultural landscapes have remained relatively stable for millennia because of a peculiar geographic circumstance, for example, the former annual deposition of sediment in the Nile River Valley (Troeh et al., 1999).

Modern agricultural landscapes are characterized by a high degree of human-managed disturbance and maintenance (Forman and Godron, 1986). More recently the conversion of natural to agricultural landscapes has included socioeconomic pressures such as social controls (described in Chapter 7, Flora, 2004) in addition to localized pressures for an increase in food supply. As pressures have increased to develop more farmland, the conversion of natural to agricultural landscapes has occurred on lands that are only marginally suited to intensive agricultural practices. If human intervention (i.e., management and energy) is removed, agricultural landscapes revert through succession to more natural self-sustaining ecosystems. To illustrate, the government of Brazil has converted natural forest to agriculture through the development of infrastructure such as road paving, river channeling, and energy production (Soares-Filho et al., 2001). Reversions from agricultural to nonagricultural landscapes have also occurred in areas where fields were steep, soils were infertile, and fields were a long distance from towns.

Marginal lands were converted to agriculture in the Great Plains of the United States in the 1970s. High commodity prices inadvertently encouraged the conversion of grasslands that were considered marginal for crop production into wheat (*Triticum aestivum* L.) (Eckholm, 1976; Sampson, 1981). Soil degradation occurred while these lands were managed as agricultural landscapes. If crop subsidies were removed and commodity prices remained low, these landscapes would not be economically suited for cropland.

A government program in the early 1960s in the former Soviet Union illustrates a misguided attempt to manage agricultural landscapes with the primary criterion being the national goal of increased food production. The program targeted grasslands in northern Kazakhstan, western Siberia, and eastern Russia for cropland expansion. The legacy of the program was 17 million hectares of degraded land, four million of which were totally lost to production in a span of 3 yr, 1962 to 1965 (Eckholm, 1976).

The natural landscapes that replace abandoned agricultural landscapes may not be equivalent to the original landscape because of permanent changes to the environment. In the past, agroecosystem sustainability was often assumed to be synonymous with the protection of crops from natural hazards by subjugating nature rather than by working with nature (Ryszkowski, and Jankowiak, 2002; Chapter 7, Flora, 2004). In either case substantial human labor and energy expenditure are needed to sustain an agricultural landscape. Sustainable agricultural landscapes require proper planning, design, implementation, and maintenance. Agricultural landscapes without some form of maintenance quickly become dysfunctional. Conservation practices within the agricultural landscape are first implemented at the RLU scale. Changes to the landscape as a whole occur when these practices are implemented on a threshold number of RLUs.

The discipline of soil and water conservation has long advocated specific practices designed to minimize natural resource degradation within agricultural settings. In the past these practices have been primarily applied to fields without consideration of the RLU or the wider landscape. The discipline of landscape ecology emphasizes the importance of understanding the interactions (energy, material, and biotic flows) among elements within a landscape. Such an understanding allows us to better understand how a simple localized change in management of a cropped field can dramatically influence a broad geographical area when this practice is replicated many times within a landscape. As an example, the draining of a temporary wetland of perhaps 10 m^2 within a field might appear to have a minimal impact on a particular landscape. However, when repeated many times, significant degradation in the hydrological functioning and habitat value of the landscape can occur.

When soil conservation and management practices are applied to many fields in a landscape the emergent properties of the agricultural landscape may differ from those expected for the field or farm. For example, drainage systems designed to prevent overland flow of water (to reduce erosion) and water accumulation in toe-slope positions (to reduce field flooding) can result in increased peak stream flow and increased transitory stream flooding. In the above example the inclusion of a system of wetlands to slow water movement into the stream would be a beneficial conservation practice on the level of the landscape. However, restoring only a few wetlands on a single farm representing a tiny fraction of the watershed may be insufficient to affect stream hydrology.

Conservation practices applied to the level of the landscape require additional evaluation and consideration beyond the hierarchical level of the field and farm (Chapter 10, Francis et al., 2004). We define conservation within the agricultural landscape to be the planned, organized management, use, protection, and maintenance of agricultural landscapes with the objective of supporting the essential physical, chemical, biological, and cultural–social functions of a sustainable agroecosystem.

We begin with an analytical framework with four strategies for conservation within the agricultural landscape: stabilization, restoration, maintenance, and redefinition. We conclude with an example of an RLU from the Prairie Pothole Region of South Dakota.

The four strategies may be applied singly or in sequence, but it is important to understand the overall goal of conservation in a given landscape and the consequences of each strategy. The strategies are applied after an examination of each landscape function, the history of change in the function's activity, and the interaction with other landscape functions (Fig. 8–1). Examples of agricultural landscape functions include crop productivity, nutrient cycling, groundwater recharge, habitat provision, and waste disposal.

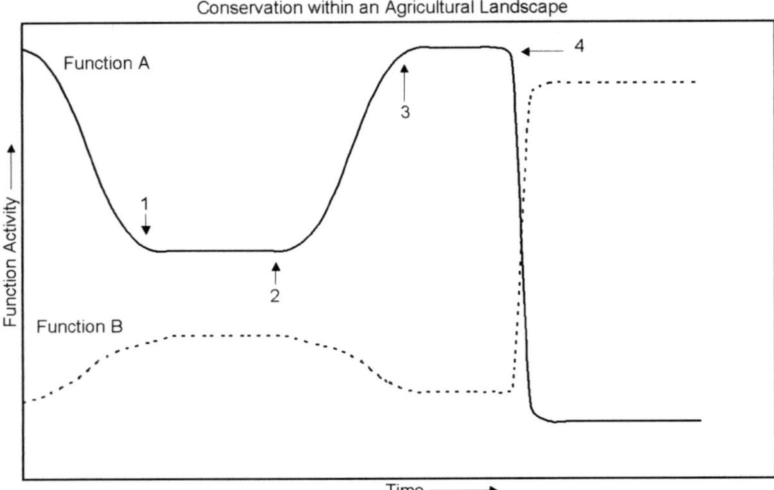

Fig. 8–1. Conservation strategies applied to two hypothetical landscape functions. The arrows indicate application of the various strategies of stabilization (1), restoration (2), maintenance (3), and redefinition (4).

CONSERVATION WITHIN AGRICULTURAL LANDSCAPES

Landscape functions that could be used to evaluate strategies for conservation within a landscape are illustrated in Fig. 8–1. Two hypothetical functions, A (wetland water storage) and B (crop productivity), were chosen to demonstrate how interactions between functions within a landscape may lead to development of different strategies. An evaluation of multiple functions and the application of various strategies is likely to demonstrate a need to optimize the entire system of landscape functions rather than optimizing only one function of the landscape.

Four strategies are shown in the diagram. The conservation strategy labeled 1 represents management designed to stop degradation of Function A (e.g., no-till to stop erosion and deposition in the wetland). Strategy 2 demonstrates the effect of restoring Function A, perhaps moving sediments from wetland to upland. The effects of conservation strategies designed to optimize Function A result in unintended consequences on Function B. Maintenance is applied when desired restoration has been achieved and is shown as Strategy 3 in the diagram. Strategy 4 in the diagram illustrates the effects of landscape redefinition, such as wetland drainage. In our hypothetical example, Function B becomes the preferred function while Function A becomes minimized. We examine each of the four strategies in more detail in the following sections.

Stabilization

Degradation of rural landscapes includes declines in soil, water, and air quality; biodiversity; and viable rural communities. The cause of landscape degradation takes many forms, and it is sometimes difficult to identify one specific cause (Barrow, 1991). Indeed, multiple factors may be functioning at different levels within the agroecosystem. A factor at the field, farm, or RLU level may be interacting with another factor acting at a higher level, at the landscape, regional, or subcontinental level, for example. In such cases, stopping landscape degradation by stabilizing a landscape function becomes more difficult than simply correcting the problem in one or multiple fields within a landscape. To reverse landscape degradation, a multifaceted approach may be needed that includes action on a number of levels of the agroecosystem spatial hierarchy, as described in Chapter 10 by Francis et al. (2004).

Stabilization prevents further decline of a function's activity, assuming the function has been in decline. For example, to stop further loss in the landscape function of crop productivity, eroded hills in cultivated landscapes could be maintained with conservation tillage practices such as chisel plowing or no-till rather than moldboard plowing (Schumacher et al., 1999). Although stabilizing a function by stopping degradation appears at first to be a simple procedure, further analysis demonstrates more complex considerations.

The strategy of stabilization is appropriate if there has been continuing decline in the targeted function activity. An assessment must be made of the baseline activity of the function. This involves some knowledge of how the targeted function has performed and changed in the past. If the time assessment indicates that the functional activity declined, but then stabilized, or if the functional activity has rebounded, then stabilization may not be needed. For example, if loss of additional soil will not further reduce crop productivity, as might happen in a deep loess soil, then conversion of the tillage system to a high residue system will not impact the crop production function within the RLU. This has been observed in soil erosion–productivity studies on loess-derived soils in Illinois and surrounding states (Olson, 1999; Schumacher et al., 1994). Conversely, the use of a high residue system such as no-till can result in a depression of the crop production function in the first few years after establishment. However, studies have shown that crop production in such a system gradually improves with time and may equal or exceed the crop production function of a tilled system in the course of a decade (Hussain et al., 1999; Rhoton, 2000). The reversion to a tilled system (i.e., stabilization of the crop production function) in this case could be counterproductive when viewed over a sufficiently long time.

Targeting of one landscape function for optimization may create other problems. In our example, even though the change in tillage system prevents further declines in crop productivity, it may also affect other landscape functions. For example, a change to conservation tillage may prevent the movement of P to surface waters by reducing sediment transport from plowed fields. However, this one change (conservation tillage) would probably not stop the decline in biodiversity in agricultural landscapes, or the decline of rural communities, or correct losses in crop productivity resulting from past erosion. In another example, the imposition of moldboard plow tillage on a no-till field to correct a perceived or real loss in the crop production function could result in a significant decline in the sediment-reducing function of the landscape. Knowledge of critical functions, the interactions of these functions, and the priority of these functions within the RLU are essential keys to optimizing the entire system of landscape functions.

Conservation practices that are based on individual fields rather than RLUs will be ineffective as landscape conservation tools. Conservation planning in agricultural landscapes must include consideration of the effects of the conservation practice on not only the matrix (agricultural fields), but also the impact on the corridors and patches within the landscape.

Conservation practices implemented at the field level through decisions made at the farm level can impact a landscape if repeated in many fields and farms. It is easier to envision this being accomplished at the level of the farm by a single owner than on the level of the landscape, especially if the landscape is composed of many farm units, unless there is outside incentive to pursue these changes on a broad scale. This is where interactions among the RLU, the landscape, and the regional and national scales become important. Programs designed to implement conservation practices on a scale wider than the individual farm must involve regional or national policy. An example is Conservation Compliance in the 1985 Farm Bill. The program targets currently eroding fields and provides incentives to convert to less erosive farming systems. The use of incentives and the targeting of this program to eroding landscapes increases the probability that the field implementation of conservation practices to stop erosion will take place on the landscape scale and in the appropriate places.

Restoration

Restoration also requires information on the history of a landscape. The term restoration implies returning the landscape to a prior state. This does not necessarily mean returning the landscape to its original natural state. Current agricultural landscapes sometimes result from the conversion (redefinition) of a landscape that was itself converted from a natural landscape. An example might be the conversion of native grassland to an introduced-grass pasture to cropland. Restoration in this instance may involve restoring the intermediate landscape (pasture) lost when the current landscape (cropland) was introduced or it may involve restoration of the original native landscape (native grassland). Restoration may also involve improving a function of the landscape that is currently present, but in a degraded state.

When the U.S. Midwest was first settled, wetland-dominated landscapes were often drained for agricultural use resulting in landscapes devoid of wetlands. Wetland restoration in these agricultural landscapes could be undertaken. In South Dakota and North Dakota settlement occurred later than in other parts of the Midwest, artificial drainage was not used to the same extent because precipitation and land values were lower. As a result many of the agricultural landscapes in this region, the Prairie Pothole Region, contain functioning wetlands.

Wetland functions fall into three broad categories relating to hydrology, biogeochemistry, and habitat/food webs. Wetland function degradation results from sedimentation and movement of agricultural chemicals into the wetlands. Function can be restored by removing sediment and recovering landscape hydrology that may have been altered by drainage. Agricultural landscapes with degraded wetlands also include eroded cropland. The Prairie Pothole Region of South Dakota includes many examples where sediment from eroded knolls in hummocky topography has degraded the wetland while the upland cropland has lost productivity through loss of topsoil. Productivity losses from erosion are persistent in landscapes where the rooting depth is limited by a dense glacial till in the C horizon (Olson et al., 1999). Loss of topsoil in these landscapes results in less available water storage for crop growth because of a reduction in root zone volume (root depth is decreased as the depth to the dense C horizon is reduced), a reduction in organic matter (organic matter improves water storage capacity of the soil), and an increase in the tendency for runoff to occur because of poor soil surface structure. Very little can be done to compensate for a loss of root zone water storage capacity in dryland agricultural environments. Soil rehabilitation practices have used soil amendments such as manure, deep tillage such as subsoiling, and the use of deep-rooted perennial crops, but these have all met with limited success. One approach would be to apply sediment removed from the degraded wetland to the most eroded parts of the landscape.

This last option illustrates the advantages of simultaneously examining all functions of a landscape. Restoring functions in an agricultural landscape in this case results in better agricultural productivity of upland soils, filtering capacity of riparian areas and wetlands, and aquatic habitat of the stream or wetland. These improvements may lead to improved crop productivity in land best suited for crop production, increased interest in hunting, fishing and tourist industries, renewed development in rural communities and improved profitability for farm families.

Maintenance

Maintenance of landscapes is necessary for human-influenced landscapes. Agricultural landscapes are created by humans to produce food and fiber. The maintenance of cropland requires labor, energy, and time inputs by humans. Without maintenance, land degradation will inevitably occur and the landscape function of food and fiber production will be impaired. Similarly the parts of the landscape that we normally think of as natural depend on conservation management within landscapes that are heavily influenced by humans.

Maintaining function of an agricultural landscape can occur after function is restored, or before it is degraded. Maintenance of a landscape requires careful monitoring of indices that measure landscape functions. Deciding on suitable function indicators is critical for a successful monitoring program. One example of a suite of indicators being developed to monitor soil processes in fields are simple measurements of soil quality (Bredja et al., 2000; Sarrantonio et al., 1996). Soil quality toolkits are available with manuals and interpretative guides (USDA-NRCS Staff, 1999). These kits are inexpensive and relatively easy to use in field situations. Monitoring of fields within landscapes using these tools is conducted over a period of time to evaluate whether the fields are aggrading or degrading in soil quality. Simple field methods for monitoring other functions within landscapes are needed. Whatever methods are used for monitoring they are of little value without a system to maintain records of observations over time.

Redefinition

Land use and therefore functions may change within landscapes. Extreme examples include the conversion of agricultural and natural landscapes into urban and suburban landscapes. Land that was valued for crop or livestock production may become valued for its ability to support houses, roads, and sewage lines and less valued for its ability to support plant growth. As an example, urban and suburban construction techniques are such that considerable subsoil compaction occurs on the land during the period of construction. The extent of the compaction often determines the extent to which future tree and shrub growth is slowed and survival rates reduced. In an urban environment, the plant productivity function of the landscape is much reduced compared with an agricultural landscape. Hydrologic functions such as groundwater recharge and retention of runoff are altered in an urban landscape compared with an agricultural landscape.

All agricultural landscapes were originally redefined in function from the prior landscape, which was usually natural. This often results in conflict between the agricultural functions and the former natural functions of the landscape. For example, draining wetlands eliminates nutrient filtering and floodwater storage functions of wetlands that are important downstream.

The danger in redefining a natural landscape to an agricultural landscape is that valuable aspects of the natural landscape that are not obviously related to agriculture will be discounted or hidden to such a degree that they become invisible until they surface as a problem. A great challenge to agroecology is to devise methods of optimizing agricultural and ecological goods and services functions of the landscape without seriously compromising the environment and economics of human-managed areas.

PRAIRIE POTHOLE REGION EXAMPLE

The Prairie Pothole Region (PPR) of North America extends across parts of North Dakota, South Dakota, Iowa, Minnesota, and several Canadian provinces. The area is dotted with glacier-formed wetlands. In the USA, the PPR is predominantly agricultural land, and in South Dakota an estimated 65% of wetland basin hectares are located adjacent to farmed fields (Johnson and Higgins, 1997).

As cropped area has increased, many wetland hectares have been drained. For example, in Iowa wetland drainage estimates are as high as 95%, compared with 35% in South Dakota (Tiner, 1984). Drainage for agricultural purposes in the PPR has impacted almost all landscape-scale functions attributed to wetlands.

Increases in tillage for crop production have accelerated erosion rates and allowed runoff to contaminate surface water. Leaching of nitrates (Barari et al., 1988) and pesticides such as atrazine (6-chloro-N^2-ethyl-N^4-isopropyl-1,3,5-triazine-2,4-diamine) is a well-documented problem in groundwater in the PPR. Sustaining the agroecosystem of the PPR includes the enhancement of wildlife populations and habitat, the improvement of surface and groundwater quality, the return of soil productivity and simultaneous restoration and maintenance of farm profitability and rural vitality.

A major step toward landscape conservation in the PPR occurred in the 1985 Farm Bill with the introduction of Conservation Compliance, the Conservation Reserve Program (CRP), and Swampbuster. This stimulated a flurry of research to define and measure properties such as erosion rates, soil quality, productivity, hydric soils, and wetland borders. Conservation tillage practices replaced moldboard plowing for many farmers who elected to participate in the federal program. In South Dakota nearly 10% of agricultural lands, including many acres of wetlands, were enrolled in the Conservation Reserve Program. The Farm Bill included several approaches to landscape conservation.

Conservation Compliance was designed to stabilize soil and water resources by reducing soil erosion and runoff from cropped fields. Swampbuster used a similar approach by limiting the expansion of wetland drainage for agricultural purposes. The Conservation Reserve Program took a further step and focused on restoring function. An attempt was made to convert highly erodible agricultural lands to grassland or woodland ecosystems by supporting the planting of perennials. Later programs helped to maintain the functions restored by extending CRP contracts and providing incentives for creating permanent buffer strips.

Common signs of wetland degradation are loss of water storage capacity, increased nutrient load, loss of biodiversity, and decreases in wildlife abundance and habitat. Slowing wetland degradation in the PPR can be achieved by limiting cultivation during dry years when seasonal or temporary wetland basins are dry and by reducing pesticide applications to wetlands for the purpose of reducing "nuisance" species (e.g., mosquitoes, blackbirds). Stabilization may also include ceasing additional drainage with new tiling systems or surface drainage trenches. Although sedimentation is a natural process in PPR wetlands (which are generally closed basins), wetland degradation occurs from excessive sedimentation originating from tilled fields. Using conservation tillage for upland acres can slow degradation due to sedimentation.

Restoring wetland function is often assumed to accompany "wetland restoration." Some indexes of success have been investigated for habitat quality and biodiversity (Juni, 2001). The restoration of wetlands has been accomplished through several federal and private programs. A common approach to restoration begins with the removal of drainage mechanisms. Wetland plants may be reestablished through plantings and/or soil seed banks. Once hydrology and vegetation are restored, wildlife habitat begins to develop, and an increase in species diversity usually follows. Generally wetland restoration has been preferred to wetland creation. In cases where wetlands have been restored, maintenance is less difficult than those where they have been created. Wetland mitigation (the trading of wetland drainage for wetland restoration or creation) has been a legal and/or political tool used to limit wetland loss, but allow wetland conversion.

The fourth approach to landscape conservation is redefinition. Many wetlands have been converted for urban development and agricultural production. Recently the loss of prior landscape function has been recognized as the cause of unintended problems in the redefined landscape. For example, agricultural conversion of wetlands has been shown to increase the frequency and severity of stream flooding (e.g., Vermillion River in South Dakota and Red River in North Dakota). Conversion of marginal agricultural landscapes to wetland and wildlife habitat landscapes is an example of the application of redefinition of an agricultural landscape that may improve functioning of adjacent agricultural landscapes within a region.

The approach used for landscape conservation is often linked to mechanisms of social control (voluntary or involuntary), which are discussed in Chapter 7

(Flora, 2004). Landowners who view wetlands as a part of the landscape have internalized wetland conservation. Farmers choosing to participate in farm programs are seeking the positive economic sanctions (voluntary) of certain program benefits and simultaneously realizing that there will be negative economic sanctions (involuntary) if certain conservation practices are ignored.

SOUTH DAKOTA CASE STUDY

In 1985, research studies began to investigate economic and agronomic interactions of farming systems in eastern South Dakota. Both on-farm and experiment station trials were conducted. The farm used in this case study was part of a larger farming systems study and is located in the Prairie Pothole Region of South Dakota, where a predominant landscape use conflict occurs between crop production and wetland preservation. Figure 8–2 shows the hummocky PPR terrain of the case study farm. The major agricultural land uses in the region include corn (*Zea mays* L.) and soybean [*Glycine max* (L.) Merr.] production, with some small grains, hay, and pasture. The case study farm consists of approximately 340 ha, with 252 cropland hectares and 36 wetland hectares. According to USDA Land Capability Class/Subclass, 7% of the land has no limitation for crop production (Class 1), 43% is limited by erosion (2e) and root zone limitation (2s), 22% is limited by wetness (2w), and 26% is severely limited by erosion (3e and 4e). The crop rotation is typically alfalfa (*Medicago sativa* L.)–oat (*Avena sativa* L.) for the first year, alfalfa (1–4 yr), soybean, and then corn. The farming system has been certified as organic, and thus uses no synthetic pesticides or fertilizers. Livestock operations include swine (*Sus scrofa*) and beef (*Bos taurus*) production on a regular basis and poultry production occasionally. It is a family-owned farm. Initial research showed that the farm as a whole was profitable and environmentally friendly (Smolik et al., 1994). However, one area chronically had negative net returns and included wetlands with high concentrations of P and low plant species diversity. The croplands and wetlands examined constituted an RLU for the larger landscape.

Bartels and Anderson (1993) listed nine steps in the process of wetland restoration that can be adapted for landscape restoration in this case. The steps are:

1. Identify the problem.
2. Determine the objectives.
3. Inventory the resources.
4. Analyze the resource data.
5. Formulate (feasible) alternative solutions.
6. Evaluate alternative solutions.
7. Client determines a course of action.
8. Client implements the plan.
9. Evaluation of the results of the plan.

Using this process, the problems were identified as negative net return, high P loads in the wetland, and low species diversity. The objectives clearly followed: positive net return, reduced P load, and increased species diversity. The next steps were to

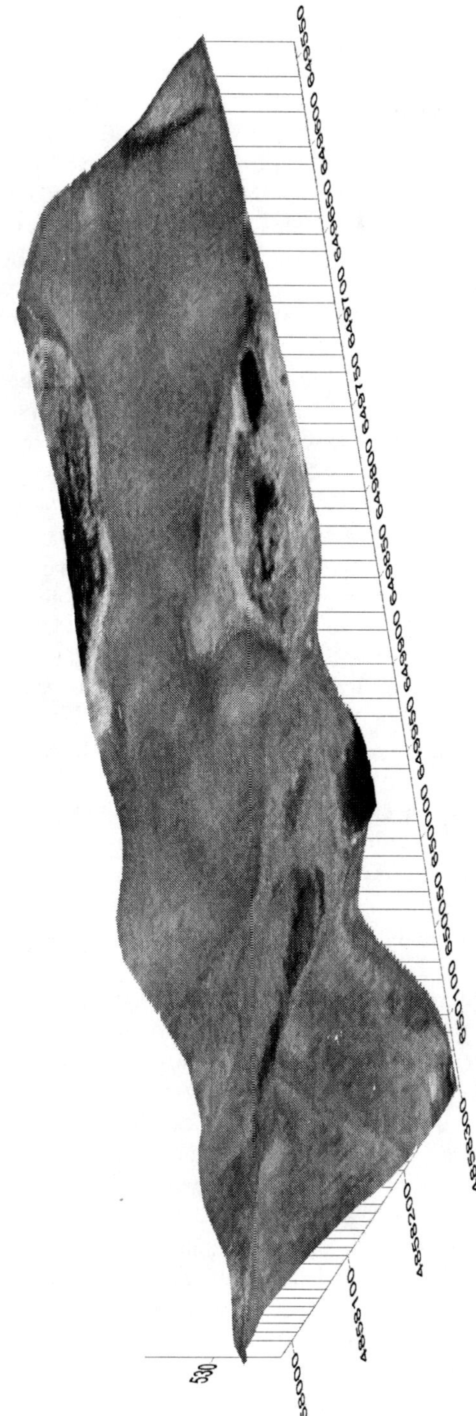

Fig. 8–2. Aerial photograph of the Prairie Pothole Region case study area in eastern South Dakota. The dark low-lying areas are wetlands within the landscape.

spatially inventory field resources using global positioning systems and geographical information systems technology and to analyze the resource data.

Nearly 30% of the RLU was classified as wetland hectares, with 22% of those being cropped. For crop yield analysis, the RLU was divided into zones: wetland zone, wetland edge extending out to 25 m zone, 25- to 50-m zone, and field zone. Four-year-average soybean yields from 1991 to 1995 were 0 Mg ha^{-1} in cropped wetlands, 2.05 Mg ha^{-1} in the zone extending from the wetland edge to 25 m, and 2.15 Mg ha^{-1} from 25 to 50 m. Corn yields were 0 Mg ha^{-1} in the cropped wetland area, 0.44 Mg ha^{-1} in the wetland edge to 25 m zone, and 4.39 Mg ha^{-1} in the 25- to 50-m zone. At the RLU scale, a composite yield was determined as the average of the yield in each zone weighted by the proportion of RLU hectares in each zone, including the portion of RLU hectares farmed within the wetland where no crop production occurred. At the RLU scale (110 ha), net returns per hectare were negative $21 for corn and positive $203 for soybean (with organic premiums). Soil P concentrations were 27.3 ppm in the wetland, 4.8 in the zone extending 25 m from the wetland edge, and 4.4 ppm in the 25- to 50-m zone. Emergent wetland vegetation included 32 species, with a predominance of cattail (*Typha* spp.).

The hummocky character of this landscape (Fig. 8–2) has a high potential for the twin processes of water and tillage erosion. These are separate processes that interact to result in a characteristic pattern of soil loss and deposition within the landscape. Detailed elevation measurements using a survey-grade GPS were combined with models of tillage erosion (Lindstrom et al., 2000) and water erosion (van Oost et al., 2000) to illustrate the soil loss and deposition patterns occurring within the studied RLU. Figure 8–3 (Tillage Erosion) illustrates the impact of tillage erosion alone on soil loss and deposition within this RLU. Soil loss from tillage erosion occurs within the landscape on the convex portions of the landscape. Tillage erosion rates are most severe on convex slopes that have steep changes in slope gradient (shoulders and the crests of knolls). This is different from erosion patterns observed with water erosion, where the most severe erosion rates occur in linear backslopes with high slope gradients. The impact of water erosion on soil loss and deposition within this RLU is illustrated in Fig. 8–3 (Water Erosion). Note the different erosion patterns occurring in the RLU from the separate erosion processes.

Both water and tillage acted in concert to move soil downslope. A net movement of soil occurred from crests and shoulders due to tillage erosion onto the lower backslope where water erosion continued the soil movement downslope and into the wetlands (Fig. 8–3, Tillage and Water Erosion). The combination of water and tillage erosion results in a degradation of the productivity function in the cropped area by reducing root zone depth on the crests and shoulders (Schumacher et al., 1999). Soil lost from the crests and shoulders as a result of tillage erosion and from the lower backslopes because of water erosion is ultimately deposited in the wetland, producing severe degradation of the wetland.

Options for change considered were wetland drainage, changes in crop selection, enrollment in government conservation programs, and use of buffer strips around the wetlands. In 1995, the owner decided to change management in the RLU to enhance crop productivity and wetland function by establishing buffer zones around each wetland without enrollment in a government program. Three species were planted in the buffer strips: smooth bromegrass (*Bromus inermis* Leyss.), or-

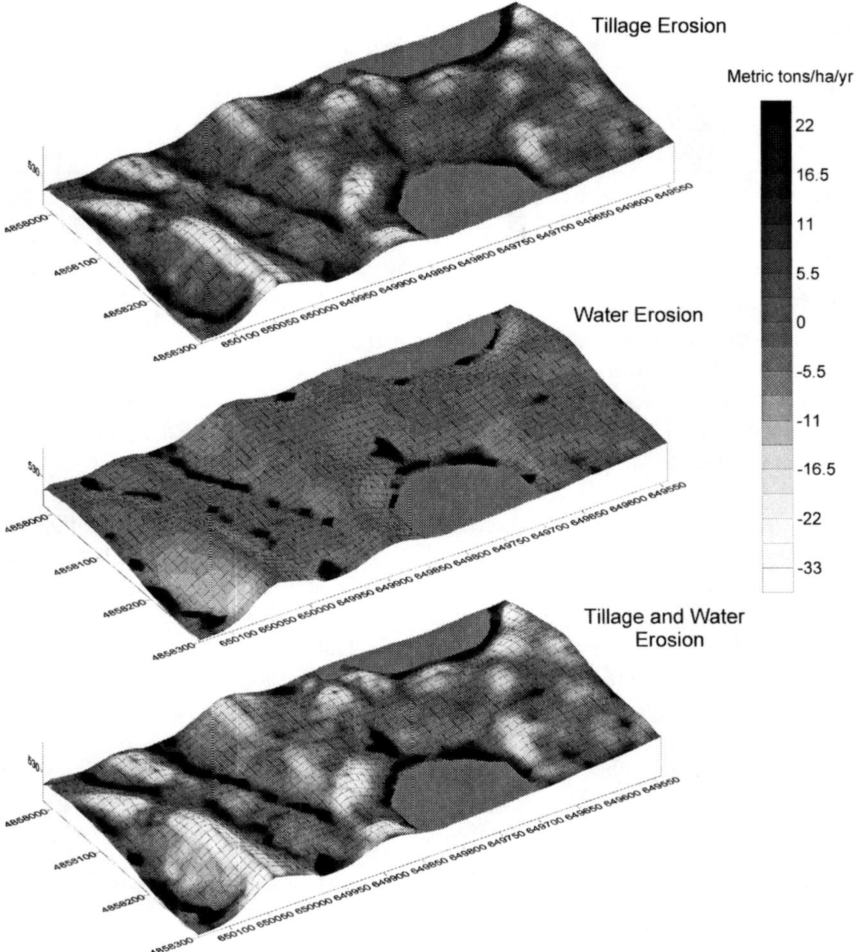

Fig. 8–3. Maps showing the application of tillage and water erosion models applied to the Prairie Pothole Region case study area in eastern South Dakota. The models simulate soil loss and deposition within the landscape. Soil loss in the scale is given by negative numbers (light areas in the graph) and soil gain by positive numbers (dark areas in the graph). Sedimentation is modeled to the edge of the respective wetlands. Sediment distribution within the wetland was not modeled.

chardgrass (*Dactylus glomerata* L.), and alfalfa. Composite yields in cropped hectares increased and net returns to management per hectare, averaged across the four crop species in the rotation (oat, alfalfa, soybean, and corn) increased from \$27 to \$57 ha^{-1}. At the RLU scale, because buffers reduced cropped hectares, net returns to management decreased. In Table 8–1, net returns for several federal program options are compared at both per hectare and per RLU scales. Highest net returns at both scales occurred with enrollment in the Wetland Reserve Program. The farm owner, however, did not choose to participate in a federal program. Without program support, highest net returns at the RLU scale occurred with farmed-through wetland management.

Table 8–1. Net returns of management at two scales, comparing four wetland management scenarios.

	Wetland management			
Scale	No buffer	Buffer	WRP†	CRP‡
\$ RLU^{-1}§	7460	6120	7540	7440
\$ ha^{-1}	27	57	71	70

† WRP, Wetland Reserve Program.
‡ CRP, Conservation Reserve Program.
§ RLU, repetitive landscape unit.

Table 8–2 summarizes changes in soil and water N and P concentrations. The buffers reduced nutrient loads to the wetland and accumulated them in crop or buffer areas. The buffer strip was removed for hay in mid July after the nesting season for most waterfowl species. Although only three species were planted in the buffer, the cessation of tillage allowed an additional 29 species to be inventoried, none of which was a noxious weed species.

At this point surveys were conducted in the PPR to determine farmer attitudes about wetlands. Surveys were collected at farm shows and local and state fairs in eastern South Dakota. Farmers were asked to rank the importance of eight crop production problems and eight benefits associated with wetlands. Farmers ranked inability to plant and added time to work around wetlands as the number one problems. Wildlife habitat was overwhelmingly ranked number one for wetland benefits.

Most wetlands in the PPR are located on private agricultural lands where management decisions are made by the land owner and/or farm operator. Profitability is a major objective at the field and farm scale. However, wetland functions valued by society (e.g., habitat and hydrology) often operate at regional scales. Combinations of personal attitudes of landowners and regulatory or incentive programs at the federal or regional level will likely determine the approach taken to optimize wetland function and farm profitability in the PPR. Landscape conservation will

Table 8–2. Nitrogen and P levels in soil, plant, and water across landscape positions as influenced by buffer strips.

		Landscape position			
Nutrient	Management	Wetland	Buffer	Edge–25 m	25–50 m
		Soil			
Kjeldahl-N, g kg^{-1}	buffer	3.6	3.0	2.6	2.6
Bray P, mg kg^{-1}	no buffer	18.8	10.7	6.2	7.6
		Plant			
N, kg ha^{-1}	buffer	109	47	--	--
	no buffer	140	NA	--	--
P, kg ha^{-1}	buffer	13	7	--	--
	no buffer	18	NA	--	--
		Water			
NO_3–N, mg kg^{-1}	buffer	1.15	--	--	--
	no buffer	0.93	--	--	--
Ortho-P, mg kg^{-1}	buffer	0.63	--	--	--
	no buffer	0.87	--	--	--

occur in the PPR when appropriate conservation practices as applied to RLUs are implemented throughout the landscape.

SUMMARY AND CONCLUSIONS

Conservation within agricultural landscapes is the planned, organized management, use, protection, and maintenance of agricultural landscapes, with the objective of supporting essential physical, chemical, biological, and cultural–social functions of a sustainable agroecosystem. Conservation practices must be applied to a significant number of RLUs within agricultural landscapes to meet the above objectives. A repetitive landscape unit (RLU) is defined as the minimum land area that contains landscape elements in the same proportion as the larger landscape. Four general conservation strategies may be applied to RLUs to achieve the objectives of a sustainable agroecosystem. Conservation strategies include stabilization, restoration, maintenance, and redefinition. Stabilization strategies are applied to stop degradation within an agricultural landscape, restoration strategies reverse past degradation of the agricultural landscape, maintenance strategies implement conservation practices that maintain the restored agricultural landscape, and redefinition of an agricultural landscape to another form of landscape is an extreme conservation strategy that fundamentally alters the matrix, the dominant element of the landscape.

Application of these strategies requires discernment of how conservation practices associated with these strategies affect other landscape functions. Optimization of conservation practices is critical when improvement of a desired landscape function interferes with or degrades other landscape functions. Knowledge is needed of landscape history and current landscape functions and interactions. An ongoing program to monitor and document changes in landscape functioning is an important component of conservation within agricultural landscapes. Agricultural and ecological goods and service functions sometimes come into conflict. The challenge to agroecology is to devise methods and practices that optimize these functions without seriously compromising the environment and economics of human-managed areas. Significant repetition of conservation practices throughout the landscape requires changes in policy, attitude, and a greater understanding of the importance of non-agricultural production functions in agricultural landscapes.

STUDY QUESTIONS

1. Describe a RLU typical of an agricultural landscape in your region. What are the basic landscape elements (matrix, patch, and corridor)?

2. Using the RLU you have described, design conservation plans for each of the four strategies: stabilization, restoration, maintenance, and redefinition.

3. Choose one strategy and discuss what policies and incentives would be necessary to accomplish the conservation goal. What implications would there be for rural communities?

4. Examine some examples of failed conservation efforts from around the world. What do they have in common? How can we learn from these mistakes and prevent them from happening again?

REFERENCES

Barari, A., T.C. Cowman, and D.L. Iles. 1988. Evaluation of data on nitrate concentrations in the Big Sioux Aquifer. Open-file Report 54UR. DWNR-SDGS, Vermillion, SD.

Barrow, C.J. 1991. Land degradation. Cambridge Univ. Press, Cambridge, UK.

Bartels, R.M., and M.W. Anderson. 1993. Wetland restoration considerations. p. 103–108. *In* J.K. Mitchell (ed.) Integrated Resource Management and Landscape Modification for Environmental Protection, Proc. Int. Symposium. 13–14 Dec. 1993. Chicago IL.

Brejda, John J., Thomas B. Moorman, Douglas L. Karlen, and Thanh H. Dao. 2000. Identification of regional soil quality factors and indicators: I. Central and Southern High Plains. Soil Sci. Soc. Am. J. 64:2115–2124.

Eckholm, E.P. 1976. Losing ground. Worldwatch Institute. Pergamon, Oxford, UK.

Flora, C.B. 2004. Community dynamics and social capital. p. 93–108. *In* Agroecosystems analysis. Agron. Monogr. 43. ASA, CSSA, SSSA, Madison, WI.

Forman, R.T.T., and M. Godron. 1986. Landscape ecology. John Wiley and Sons, New York.

Francis, C., G. Lieblein, S. Gliessman, T.A. Breland, N. Creamer, R. Harwood, L. Salomonsson, J. Helenius, D. Rickerl, R. Salvador, M. Wiedenhoeft, S. Simmons, P. Allen, M. Altieri, C. Flora, and R. Poincelot. 2003. Agroecology: The ecology of food systems. J. Sustain. Agric. 22(3):99–119.

Francis, C., L. Salomonsson, G. Lieblein, and J. Helenius. 2004. Serving multiple needs with rural landscapes and agricultural systems. p. 147–166. *In* Agroecosystems analysis. Agron. Monogr. 43. ASA, CSSA, SSSA, Madison, WI.

Gliessman, S.R. 1998. Agroecology: Ecological processes in agriculture. Ann Arbor Press, MI.

Gregorich, E.G., L.W. Turchenek, M.R. Carter, and D.A. Angers (ed.) 2001. Soil and environmental science dictionary. Canadian Society of Soil Science, CRC Press, Boca Raton, FL.

Hussain, I., K.R. Olson, S.A. Ebelhar. 1999. Impacts of tillage and no-till on production of maize and soybean on an eroded Illinois silt loam soil. Soil Tillage Res. 52:37–49.

Johnson, R.R., and K.F. Higgins. 1997. Wetland resources of eastern South Dakota. South Dakota State Univ., Brookings.

Juni, S. 2001. Assessment of wetland mitigation in South Dakota. M.S. thesis. South Dakota State Univ., Brookings.

Lev-Yadun, S., A. Gopher, and S. Abbo. 2000. The cradle of agriculture. Science 288:1602–1603.

Lindstrom, M.J., J.A. Schumacher, and T.E. Schumacher. 2000. TEP: A tillage erosion prediction model to calculate soil translocation rates from tillage. J. Soil Water Conserv. 55:105–108.

Lowdermilk, W.C. 1953. Conquest of the land through seven thousand years. Agric. Inf. Bull. 99. USDA-SCS, Washington, DC.

Olson, K.R., D.L. Mokma, R. Lal, T.E. Schumacher, and M.J. Lindstrom. 1999. Erosion impacts on crop yield for selected soils of the North Central United States. *In* R. Lal (ed.) Soil quality and soil erosion. CRC Press, Boca Raton, FL.

Rhoton, F.E. 2000. Influence of time on soil response to no-till practices. Soil Sci. Soc. Am. J. 64:700–709

Ryskowski, L. and J. Jankowiak. 2002. Development of agriculture and its impact on landscape functions. *In* L. Ryskowski (ed.) Landscape ecology in agroecosystems management. CRC Press, Boca Raton, FL.

Sampson, R.N. 1981. Farmland or wasteland. Rodale Press, Emmaus, PA.

Sarrantonio, M., J.W. Doran, M.A. Liebig, and J.J. Halvorson. 1996. On-farm assessment of soil quality and health. p. 83–106. *In* J.W. Doran and A.J. Jones (ed.) Methods for assessing soil quality. SSSA Spec. Publ. 49. SSSA, Madison, WI.

Schumacher, T.E., M.J. Lindstrom, D.L. Mokma, and W.W. Nelson. 1994. Corn yield: Erosion relationships of representative loess and till soils in the North Central United States. J. Soil Water Conserv. 49:77–81.

Schumacher, T.E., M.J. Lindstrom, J.A. Schumacher, and G.D. Lemme. 1999. Modeling spatial variation in productivity due to tillage and water erosion. Soil Tillage Res. 51:331–339.

Smolik, J.D., T.L. Dobbs, and D.H. Rickerl. 1994. Relative sustainability of alternative, conventional and reduced till farming systems. J. Altern. Agric. 10:25–35.

Soares-Filho, B.S., R.M. Assunção, A.E. Pantuzzo 2001. Modeling the spatial transition probabilities of landscape dynamics in an amazonian colonization frontier. Bioscience 51:1059–1067.

Tiner, R.W., Jr. 1984. Wetlands of the United States: Current status and recent trends. Nat. Wetland Inventory, U.S. Fish and Wildl. Serv., U.S. Gov. Print. Off. Washington, DC.

Troeh F.R., J.A. Hobbs, and R.L. Donahue. 1999. Soil and water conservation: Productivity and environmental protection. Prentice Hall, Upper Saddle River, NJ.

USDA-NRCS staff. 1999. Soil quality test kit guide. USDA-NRCS, Soil Quality Inst., Ames, IA.

van Oost, K., G. Govers, and P. Desmet. 2000. Evaluating the effects of changes in landscape structure on soil erosion by water and tillage. Landscape Ecol. 15:577–589.

9 Geographic Information Systems and Landscape Analysis

ROBERT M. CALDWELL

School of Natural Resources
University of Nebraska
Lincoln, Nebraska

The functional characteristics of rangeland plant species (i.e., height, biomass, specific leaf area, longevity, and litter quality) do not have continuous distributions (Walker et al., 1999). Instead, the species are organized in a small number of distinct functional groups that share characteristics. This organization increases resiliency, helping ensure ecosystem function continues under the impact of changing environmental stresses (Walker et al., 1999). In "Cross-Scale Morphology, Geometry, and Dynamics of Ecosystems," Holling (1992) developed the idea that ecosystem dynamics are dominated by a relatively small number of processes, each expressed according to a distinct temporal frequency and a discontinuous spatial pattern. He used a large data set of size measurements to demonstrate that animals have significant discontinuities in the distributions of their size classes. The same size "clumps" exist across a surprising array of contrasting environments, from boreal forest to short grass prairie, for a variety of animal types: birds vs. flying mammals, and herbivores vs. carnivores vs. omnivores. Holling presented a series of hypotheses that might explain the similarities in size-class distributions, evaluated each with the available evidence, and concluded that terrestrial organisms have evolved under the pressure of only a few dominant ecological processes. The "lumpy architecture" of size classes reflects specific adaptations to the temporal and spatial frequencies at which the processes operate. Landscapes are best viewed as an array of distinct patches subject to a limited number of natural hazards each having a distinct cycle time.

Appreciation for the patch nature of crop landscapes has been increased by the introduction of yield monitors, global positioning systems (GPS), and remote sensing. Technology now exists in the general farm population for mapping productivity of whole farms (e.g., hundreds or thousands of hectares) at resolutions down to 10 m^2 for yield and 1 m^2 for crop images. Starting in 1997, during a series of winter extension meetings in Nebraska on precision farming technologies, farmers were asked to describe their use of yield mapping in their management systems. Their motivations to collect yield maps varied. Some had a commitment to using state-of-the-art information technologies on their farm based on their understanding of the theoretical benefits of site-specific management. Others had some curiosity

 Agroecosystems Analysis, Agronomy Monograph no. 43.

about mapping and viewed the expense of the GPS and yield monitor as small relative to the purchase cost of their combine. The most common benefit expressed by farmers was the new appreciation for variations of yield within fields—something easily recognized by combine operators watching the display of the yield monitor and confirming qualitative observations and impressions of experienced farmers.

A recurring concern expressed by farmers in extension meetings was how little they used their yield maps in decision making. Yield maps occasionally revealed significant yield differences between corn (*Zea mays* L.) hybrids or soybean [*Glycine max* (L.) Merr.] cultivars planted next to each other. None of the farmers was able to confirm they made a positive return on their investment in yield mapping, and only a few were using maps from previous years to calculate application rates for fertilizer or seed.

To help farmers better utilize information on spatial variability within fields, two lines of research and extension were developed: (i) on-farm research using precision farming technology and (ii) landscape modeling for use in decision support systems. The Nebraska Soil Fertility Network resulted from the on-farm research initiative. Sites in the network are near the Nebraska boundary of the Western Corn Belt Plains (Fig. 9–1; USEPA, 2000). Farms in the transition to the Nebraska Sandhills (i.e., the northwest sites in Fig. 9–1) have soils dominated, to varying degrees, by sands (e.g., Valentine fine sand, mixed, mesic Typic Ustipsamments). Other farms contain a diversity of soil types, often silt loams (e.g., Crete silt loam, fine, montmorillonitic, mesic Pachic Argiustolls). Most, but not all, of the sites are irrigated with center-pivot irrigation equipment. The network includes experimental treatments for most soil amendments (N, P, K, S, and Mg fertilizers, lime, composted manure, and insecticide) applied using commercial production equipment (aerial application, fertilizer spreaders, and injectors attached to irrigation equipment) on corn and soybean. The crops are harvested using a variety of commercial combines

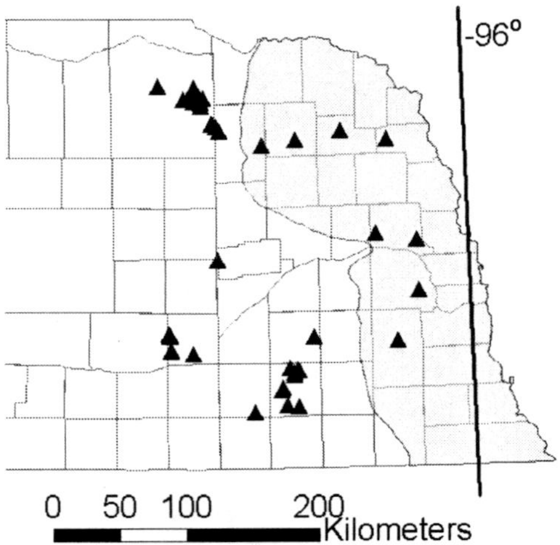

Fig. 9–1. Distribution of Nebraska Soil Fertility Network sites. The meridian shown, 96° W long, is the dividing line between Zone 14 and Zone 15 of the Universal Transverse Mercator (UTM) system. Nebraska's portion of the Western Corn Belt Plains (USEPA, 2000) is shaded.

equipped with yield monitors and GPS units. The cooperating farmers are responsible for all their field activities, from applying the treatments to mapping the experiment at harvest. The role of university staff is to facilitate farmer selection of a treatment design and an experimental design, produce the plot randomization, and analyze the data. Through the network, farmers have shared more than 4000 ha of map data.

Landscape modeling was pursued as a second direction for research and extension. One goal of this activity was to support our interpretation of the on-farm research data. A specific concern at the outset was the recognition that overland flow of runoff and intermittent ponding can have a large influence on yields from some positions on the landscape. A second goal was to enable evaluation of new management options using historical weather records as the basis for spatial risk assessment and strategy analysis. For the modeling, we adopted the object-oriented framework called JanuSys (Caldwell and Fernandez, 1998) and developed yield map simulation capabilities (Caldwell, 2003). Organization of input data using a geographic information system (GIS) was required for a number of steps in the process of yield map modeling.

During a presentation at the 1998 annual meeting of the Nebraska Sustainable Agriculture Society, the audience expressed concern about the possible unintended consequences of public support of precision farming through research and development. Following a presentation entitled "Precision agriculture and sustainable agriculture: Are they compatible?" the audience made a case that GIS, GPS, and other advanced information technologies are not scale neutral: larger farms with larger capital reserves can adopt the technology more quickly. One scenario discussed was that the technology could contribute to a loss of small farms as farmers with yield mapping gain an advantage when negotiating land rentals. Large farms that end up dependent on the technology may become less able (instead of more able) to diagnose site-specific production problems or become less able to pay attention to problems once diagnosed. In this chapter, I explore the idea that GIS technology should be a necessary part of modern agroecological research and education. After an introduction to GIS in agriculture, I present results from on-farm research that illustrate the capacity to test treatments laid out across whole fields, to analyze the multiple variables that are correlated in those fields, and to delineate boundaries around problem areas. The importance of spatial record keeping is discussed next, as well as a recommendation that all farmers, whether or not they use a computer, should adopt a simple nondigital spatial record-keeping system for their farm. Concluding comments are focused on educators. If GIS technology is to be used properly in the critical test of the system dynamics shaping agriculture, such as the systems renewal cycle proposed by Holling (1992), students of agroecology will need to be trained in GIS techniques.

INTRODUCTION TO GIS TECHNOLOGY FOR AGRICULTURE

GIS technology is being employed in all facets of agricultural research and policy support. Research topics include agroecological zonation (Caldiz et al., 2001), assessment of land use conversions (e.g., Gardi, 2001; Nizeyimana et al.,

2001), evaluation of nonpoint source pollution (Corwin et al., 1999), spatial modeling (e.g., Caldwell, 2003; Frede et al., 2002; Hartkamp et al., 1999; Lagacherie et al., 2000; Priya and Shibasaki, 2001; Tucker et al., 2000; Wu and Babcock, 1999), sustainable water management (e.g., Calera-Belmonte et al., 1999; Steinhardt and Volk, 2002) and water productivity (Ines et al., 2002), sustainable land use (Carsjens and van der Knaap, 2002), sustainable grazing management (Bellamy and Lowes, 1999), assessment of agroforestry valuations (Franco et al., 2001), disease management (Barnes et al., 1999), delineation of ecological sites (Creque et al., 1999), greenhouse gas emissions from agriculture (Matthews et al., 2000), identification of critical areas for riparian buffers (Narumalani et al., 1997), mapping regions susceptible to pesticide leaching (Gerstl, 2000), analysis of landscape-level diversity (von Arx et al., 2002), and analysis of the sustainability of cropping systems (Farrow and Winograd, 2001; Stockle, 1996). Spatial and spatiotemporal statistics as applied to agricultural research and analysis (e.g., Christakos, 2000; Crosetto et al., 2000) have become important as scientists seek to make inferences about GIS data. Research on many of the applications of mapping technology to production agriculture, including methods for soil mapping and variable rate application technology, are described in the proceedings for the 6th International Conference on Precision Agriculture (Robert, 2002).

The core definition for a GIS includes computer software for analyzing spatial data and presenting the results. GIS definitions normally include specific types of data processing, such as automated cartography and database management, and include the hardware involved in data input, output, and storage (Chou, 1997). People are also considered a necessary part of the system (Environmental Systems Research Institute, 1994), given the unstructured nature of the problems solved using GIS and the role operators must take in selecting analytical methods and parameters. Even the process of recording the position of a spatial object requires value judgments. There is no single best way to assign spatial coordinates to data, and farmers regularly must work with different coordinate systems. Latitude and longitude coordinates do not define a position by themselves. Coordinates must be referenced to a datum (i.e., a standard representation for three-dimensional shape of the earth) in order to be interpreted correctly. Many different datums have been used for published maps, in large part because of adoption of new datums as better information on earth geometry becomes available. Innumerable datums are possible. Some will work significantly better in a particular region given localized variations in the shape of the earth. Important data sources for agriculture have geographic coordinates based on the North American Datum 1927, including printed soil survey maps from the USDA-NRCS (formerly the Soil Conservation Service) and topographic maps from the USGS. Data from GPS units are set in World Geodetic System 1984, which for practical purposes matches North American Datum 1983. Another choice made by GIS operators is the projection used to present spatial data. Maps normally are displayed on two-dimensional surfaces. The projection is the algorithm that translates the three-dimensional spatial data onto the two dimensions. The Universal Transverse Mercator (UTM) projection is currently the most important for farm-level mapping. Digital orthophotography, now available for much of the United States on the Internet (e.g., Microsoft, 2001; Nebraska Department of Natural Resources, 2001), is rectified and projected in UTM. New digital soil surveys, contained in State Soil Geographic (SSURGO) databases, are provided in UTM projections to match the orthophotography. GIS technology is needed in agriculture to solve problems in combining spatial data from a variety of sources and reconciling the datums and projections used in the data. Given the change in coordinate systems between UTM zones, farmers whose land spans a zone boundary (see Fig. 9–1) are faced with a number of practical problems in designing a farm-level GIS and are likely to need help from a GIS specialist.

As part of providing yield monitoring equipment, hardware manufacturers often supply software that performs basic mapping functions, that is, reading yield and GPS data from the monitor, displaying the yield map on the computer screen, printing the map(s), and organizing the files for later retrieval. Mapping utilities, either free with the equipment or purchased for a small fee, have become popular. These utilities have few features beyond map production. Most farmers interested in analyzing their maps need features found in GIS software.

Commercial farm GIS software provides yield mapping functions with spatial analysis tools. The software gives options for (i) extracting spatial data on the basis of the user's criteria (e.g., all points where yield is less than a threshold), (ii) performing spatial interpolation to calculate a complete spatial coverage from a set of point data, and (iii) calculating new types of data from mathematical combinations of the original data layers, as in a fertilizer prescription based on soil organic matter, spring profile nitrate, and historical yield levels. Full-featured farm GIS packages can accept spatial and nonspatial data inputs from a variety of sources and create output files compatible with equipment controllers. Farm GIS software has become a repository for the types of nonspatial data associated with farm record-keeping systems. Reports summarizing whole-field or whole-farm management information can be performed with database queries using the field or farm identifier, without reference to spatial coordinates. Some of the important farm GIS software packages are built on top of commercial GIS packages produced by industry-leading GIS companies. By using the same file formats as in the commercial GIS, farm professionals are able to quickly gain access to sophisticated, state-of-the-art spatial analysis tools. A license for one of these farm GIS packages can cost thousands of dollars, not including the expense of education to learn the software.

Digital orthophotography, created and certified in the USA by the USGS, forms the foundation on which other GIS data layers can be displayed and interpreted. Orthophotos give a visual context for orienting the GIS user. Vector data displayed over the images can be evaluated for their correlation with features in the image such as field and waterway boundaries or changes in soil color. Digital orthophotos also help process new aerial photographs. Once digitized, an aerial photograph can be rectified and georeferenced by associating points in the picture with points in the orthophoto and "rubber sheeting" the picture data in image processing software so that the picture elements of the two images line up. Digital orthophotos are needed for displaying digital soil maps using, for example, a SSURGO Data Base produced by the NRCS. Along with digital orthophotography and digital soil maps, digital elevation model (DEM) data is recommended for base maps in a farm-level GIS. Digital elevation model data are available from public sources on the Internet. The data describe the landscape topography. GIS operators unfamiliar with a field often need the DEM for interpreting other map layers, and modern analytical methods use the DEM to estimate the spatial dynamics of rainfall runoff, run-on, and ponding (Caldwell, 2003).

USE OF GIS IN LANDSCAPE ANALYSIS

Data from one of the experiments in the Nebraska Soil Fertility Network are presented here to illustrate the types of analyses possible using GIS. The farmer's initial interest in this case was to test whether or not it was profitable to apply N fertilizer to soybean at the beginning-pod stage of crop development. The field (Fig. 9–2) has a center pivot irrigation system with digital controls for injecting fertilizer, so the treatment itself can be performed with little labor. Two treatments (with and without fertilizer) were applied to 0.125-radian sectors of the field in a randomized complete block design with four replicates (Fig. 9–3a). Two aerial images were acquired during the season, one true color (digitized into red, green, and blue wavebands for each square meter) and one false color on near-infrared sensitive film (digitized for near-infrared, red, and green wavebands, to 1 m^2; Fig. 9–3b). The farmer harvested the field with a yield-mapping combine (Fig. 9–3c). The yield monitor recorded measurements of grain flow at the outflow of the clear grain elevator, ground speed (to estimate distance traveled), and position estimated by GPS. The monitor was set to record the data every second. Operating speed for the combine was approximately 1.5 m s^{-1}, with a swath width of 7.2 m, for a typical harvest area of 10.8 m^2 per point. After harvest, the field was mapped for soil electrical conductivity using a Veris 3100 Soil Mapping System (Veris Technologies, Salina, KS). The Veris system records two measures: one primarily sensitive to the topsoil (nominally 0–0.3 m depth, referred to below as "shallow") and one sensitive to greater depths (0–0.9 m depth, referred to as "deep"). Operating speed for the probe averaged 4.8 m s^{-1}, with an average swath width of 15 m (72 m^2 per point). The soil probe and the combine were operated north to south in the field.

Application of N fertilizer did not increase soybean yield (Table 9–1). One immediate benefit of the study recognized by the farmer was that by running the experiment he saved one-half his anticipated fertilizer cost compared with the whole-field application he had planned for 2000.

On the basis of the map information (Fig. 9–3), the farmer was able to identify three problem spots within the field. To explore the features of the problem areas, we isolated a 6-ha area surrounding the spot most central to the field (see rectangle drawn in Fig. 9–2 and 9–3). The problem area contains a patch with high red-band reflectance (Fig. 9–3b) and low yields (Fig. 9–3c), positioned on a "saddle" region connecting the field's southwestern upland with the mideastern upland (Fig. 9–2b). For statistical analysis, observations within the 6-ha region were aggregated to match the data source with the coarsest resolution: the soil electrical conductivity measurements. To aggregate the data, the harvest area for each yield point was identified based on harvest swath width and the GPS positions prior to and after the point. Yield measured at the point was applied to each square meter in the point's harvest area. The area associated with each soil electrical conductivity measurement was then identified by classifying each square meter according to its closest conductivity measurement. Data within the area represented by a conductivity meas-

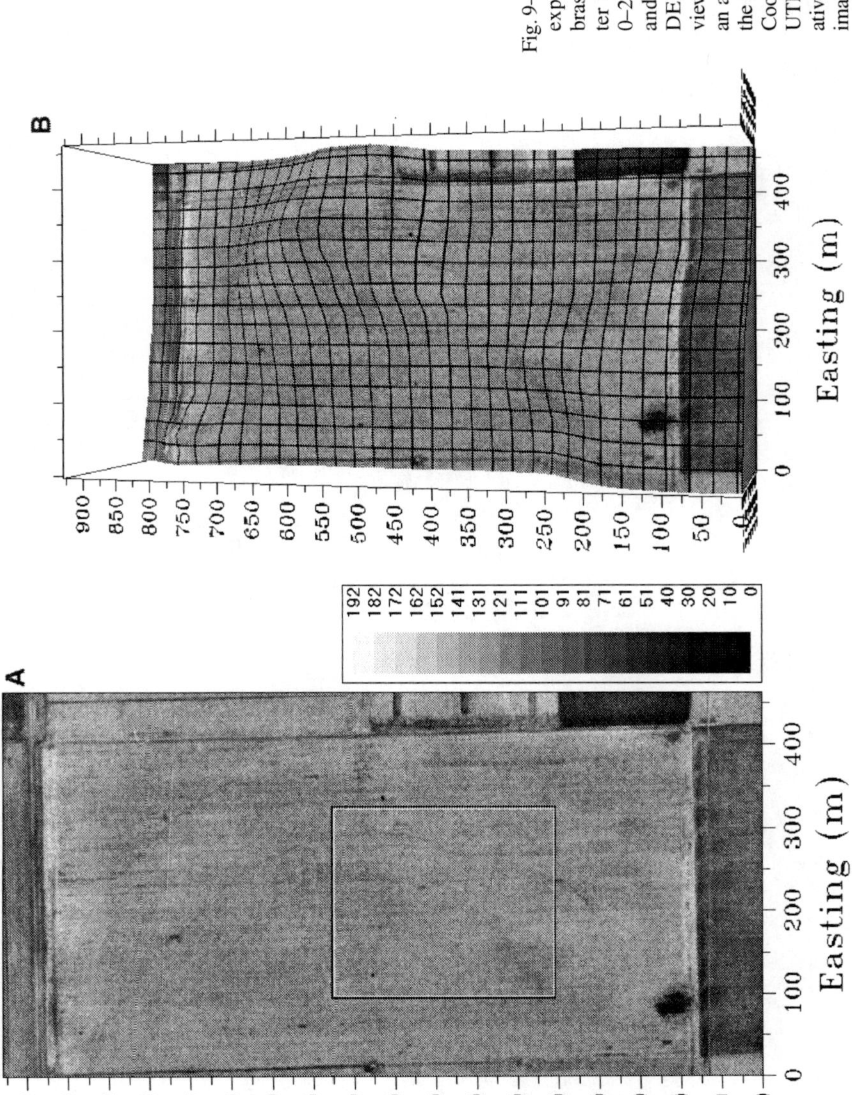

Fig. 9–2. Base map data for an on-farm experiment, Dodge County, Nebraska: (a) digital orthophoto quarter quadrangle data (scale range: 0–255 for panchromatic brightness) and (b) the image draped on its DEM data (30-m interval) and viewed in perspective. See text for an analysis of the data bounded by the rectangle shown in the figure. Coordinates, originally in Zone 14 UTM, NAD83, were translated relative to the lower left corner of the image.

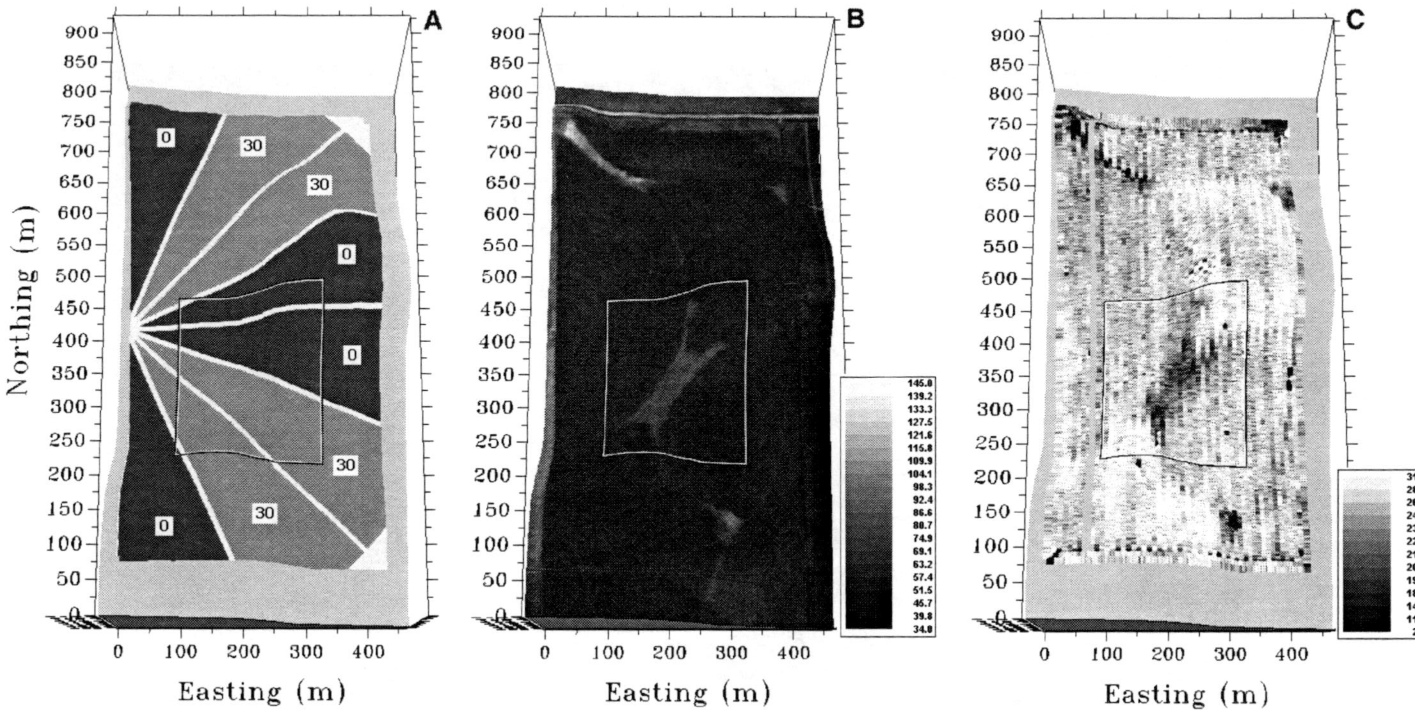

Fig. 9–3. (a) Treatment application map (0 or 30 kg N ha^{-1}), (b) red-waveband reflectance acquired 12 Sept. 2000 as the soybean crop approached physiological maturity, and (c) the yield map for an on-farm research trial. Data are draped on the DEM (compare with Fig. 9–2b). A rectangular region, shown in each map, was identified for further analysis (see text).

Table 9–1. Analysis of variance for a 30-ha on-farm experiment testing soybean response to N fertilizer. Sums of squares (SS), mean squares (MS), and hypothesis tests (Pr > F) based on F ratios are Type III (Proc GLM; SAS Institute, 1999).

Source of variation	df	SS	MS	F	Pr > F
		$\times 10^{-6}$			
Total	25 146	40.996			
Block	3	2.729	0.910	9.67	0.047
Treatment	1	0.065	0.065	0.69	0.467
Block × treatment	3	0.282	0.094		
Within-plot	25 139	38.124	0.002		

urement were averaged, resulting in 819 observations. Table 9–2 gives the correlation matrix for those data, with results from a principal components analysis (PCA; Proc Princomp; SAS Institute, 1999).

Yield was correlated with all of the variables (Table 9–2). Late-season red reflectance was the single variable most strongly correlated with yield, followed by the red-band normalized difference vegetation, which is a function of red reflectance. Soil electrical conductivity was highly correlated with yield, although much less correlated with late-season red reflectance. Among the five different types of data (i.e., yield monitor, digital elevation model, bare soil imagery, crop imagery, and soil electrical conductivity), the highest correlations were between elevation and soil electrical conductivity. Principal components analysis (Jolliffe, 2002) was used to summarize the correlations and identify a series of orthogonal axes (i.e., the eigenvectors) that can aid in interpretation of the data. The first eigenvector from the PCA represents a primarily crop-based axis, dominated by yields and crop imagery (note the absolute magnitude of the loadings in the first eigenvector, Table 9–2). The second eigenvector, accounting for less than one-half the standardized variation of the first, is a primarily soil-based axis dominated by the relationship between elevation and soil electrical conductivity. These two axes account for the agronomically important sources of information in the 6-ha area. The third principal component accounted for variation added by the early-season true-color crop image. The axis is unrelated to yield.

On 22 August and during an Extension field day on 30 Aug. 2000, the northwest corner of the field (Fig. 9–3) was scouted. At those times, a patch of severely stressed plants was visible. The plants were stunted in size and showed visible symptoms of wilt. The surface soil in the worst areas was sand. Within a few meters of those spots were areas with healthy plants growing on soils with more silt and clay in the topsoil. We inferred that the reduction in soybean growth was caused by a deficit of water possibly in combination with nutrient deficiencies, with the deficiencies being a function of the sand's physical and chemical properties. Low readings for soil electrical conductivity in the northwest corner are consistent with the presence of sand there. No observations were made inside the 6-ha area (Fig. 9–2 and 9–3; note the distance from the field entry), but we hypothesize that the same type of problems existed there as in the northwest corner.

To draw boundaries around the trouble spots, I applied a spatial aggregation method that started with the complete set of grid cells, each as their own zone, and sequentially merged neighboring zones until a small number of dissimilar zones

Table 9–2. Correlation matrix for data associated with soil electrical conductivity measurements (n = 819; critical value for $r_{\alpha=0.05}$: 0.07) within a 6-ha region of the soybean field shown in Fig. 9–3, along with the first three eigenvectors from a principal components analysis of the matrix. Eigenvectors 1, 2, and 3 accounted for 0.46, 0.19, and 0.1 proportions of the standardized variation, respectively.†

Variable		Soil electrical conductivity													Eigenvector		
name		shallow	deep	DEM	DOQQ	NIR_2	Red_2	$Green_2$	Red_1	$Green_1$	$Blue_1$	rNDVI	gNDVI	Grain yield	1	2	3
		r															
Grain yield		0.479	0.418	0.222	−0.119	0.480	−0.625	−0.580	−0.473	−0.556	−0.172	0.619	0.609	1.000	0.314	0.103	−0.020
gNDVI		0.183	0.280	−0.040	−0.020	0.859	−0.921	−0.892	−0.501	−0.497	−0.041	0.978	1.000		0.377	−0.201	0.127
rNDVI		0.209	0.313	−0.045	−0.015	0.870	−0.948	−0.843	−0.537	−0.528	−0.046	1.000			0.382	−0.186	0.123
$Blue_1$		−0.237	−0.166	−0.158	0.021	0.140	0.160	0.188	0.119	0.467	1.000				−0.086	−0.198	0.707
$Green_1$		−0.420	−0.344	−0.071	0.086	−0.327	0.591	0.523	0.625	1.000					−0.291	−0.104	0.442
Red_1		−0.374	−0.385	−0.090	0.076	−0.401	0.552	0.460	1.000						−0.280	−0.072	0.118
$Green_2$		−0.100	−0.174	0.073	0.039	−0.536	0.926	1.000							−0.344	0.205	0.115
Red_2		−0.173	−0.268	0.071	0.027	−0.668	1.000								−0.375	0.185	0.056
NIR_2		0.220	0.318	0.016	−0.003	1.000									0.312	−0.146	0.373
DOQQ		−0.122	−0.143	−0.145	1.000										−0.038	−0.145	−0.033
DEM		0.648	0.675	1.000											0.065	0.521	0.216
Soil elect.	deep	0.829	1.000												0.211	0.463	0.218
conduc.	shallow	1.000													0.189	0.505	0.093

† DEM: elevation; DOQQ: bare-soil brightness, from public digital orthophoto quarter quadrangle data; Red_1, $Green_1$, and $Blue_1$: red, green, and blue waveband reflectances, respectively, on 8 Aug. 2000; NIR_2, Red_2, and $Green_2$: near infrared, red, and green waveband reflectances, respectively, on 12 Sept. 2000; rNDVI = $(NIR_2 - Red_2)/(NIR_2 + Red_2)$; gNDVI = $(NIR_2 - Green_2)/(NIR_2 + Green_2)$.

were obtained. At each stage, every pair of neighboring zones was checked to see which pair, if merged, would add the smallest amount to the pooled within-zone variance. That pair was aggregated and the sequence continued. The probability of an error at each stage in the merger sequence was estimated using an F ratio of the change in within-zone variance (1 df) divided by the within-zone variance at that stage. By comparing that probability with a predefined α level, I created a statistical approach for identifying the number of significant zones; that is, the last stage before the estimated error was less than α. Figure 9–4 gives the results for the 6-ha region identified in Fig. 9–2 and 9–3 and analyzed in Table 9–2. Aggregations were performed on three different values to compare the sizes and shapes of the resulting zones. Two of the values were single variables: yield and the deep reading for soil electrical conductivity. The third aggregation was performed on the imagery data. Principal components analysis was used to summarize all the imagery variables (Table 9–2) into a single score, taken from the first principal component, which was then applied to the aggregation process. An α level of 0.05 was used to identify the number of significant zones: 12 for soil electrical conductivity, 10 for canopy reflectance score, and seven for yield (Fig. 9–4). The zones are mapped in Fig. 9–5. The three zonation maps each show a basic pattern that falls on the saddle region of the topography (Fig. 9–5a) and extends at that elevation onto the hill slopes. The 90-degree bends in boundaries for yield and soil electrical conductivity are a function of the resolution of the original data. One of the zones for yield,

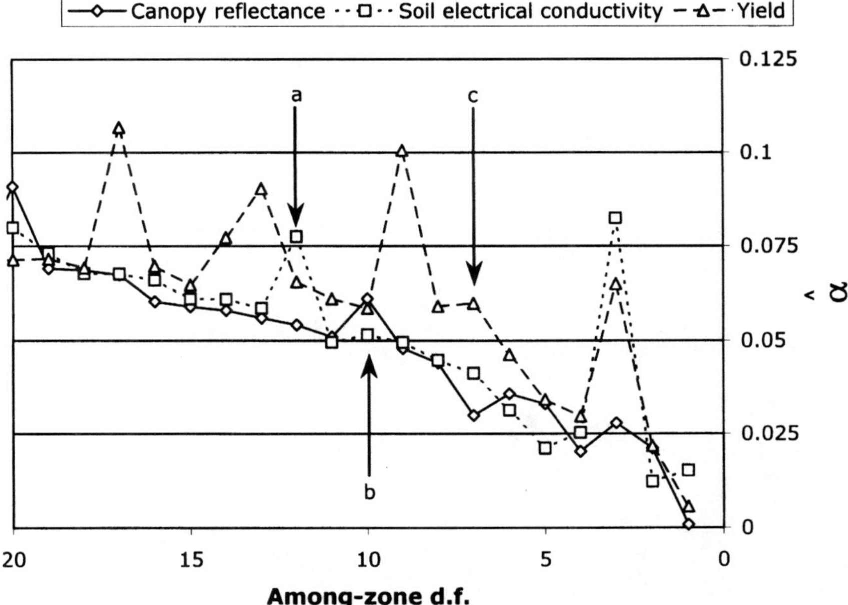

Fig. 9–4. Estimated probability $\hat{\alpha}$ of making an error when merging neighboring zones, for the final stages in the spatial aggregation sequence (left to right). The original zones were cells in a 3-m grid of the rectangular area shown in Fig. 9–2 and 9–3 ($n = 6675$). Arrows indicate the final mergers ($\alpha = 0.05$) based on zonations for (a) soil electrical conductivity (from deep-reading probe), (b) canopy reflectance data (first principal components score for crop imagery variables shown in Table 9–2), and (c) yield.

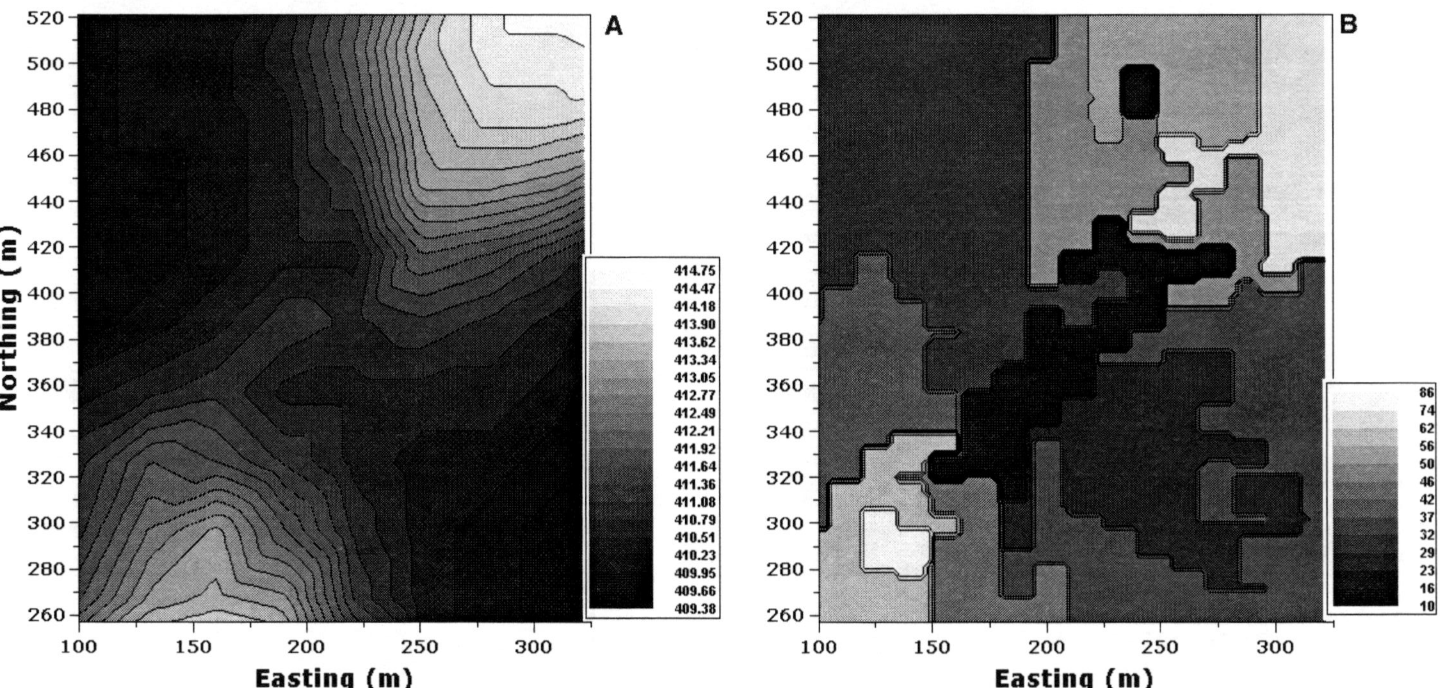

Fig. 9–5. (a) Digital elevation model (unit: m) and zonation results for (b) soil electrical conductivity (deep reading; mS m^{-1}).

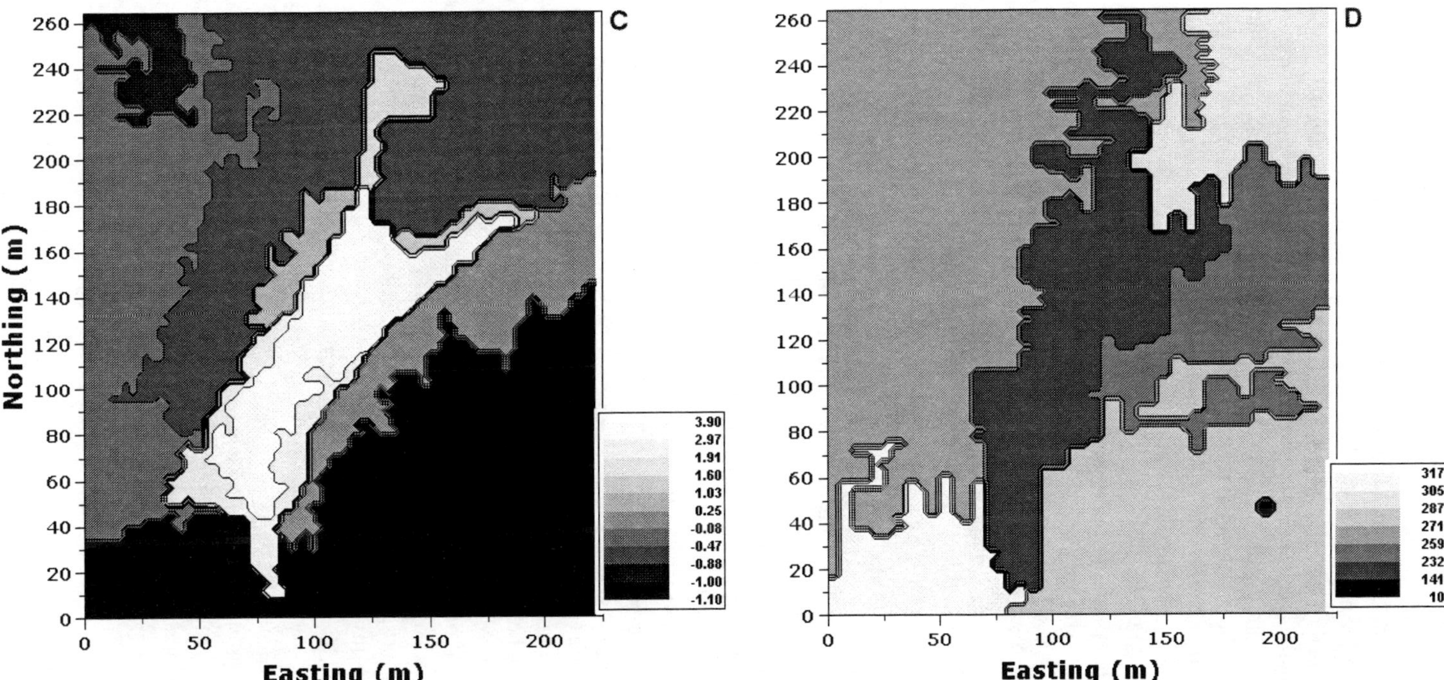

Fig. 9–5 continued. (c) canopy reflectance (first principal components score from two images each with three wavebands; unitless), and (d) soybean yield (g m^{-2}).

identified as a spot in the southeast sector of the area, is an artifact of grain flow dynamics within the combine. At that spot the operator was forced to slow to a stop and then accelerate to resume harvest. Upon accelerating, the clean grain elevator was empty and therefore the yield estimate (i.e., grain flow divided by the area harvested) dropped to essentially zero.

The analysis of the 6-ha region illustrates both landscape gradients and landscape discontinuities in which the land is divided into qualitatively different zones (Fig. 9–5). At the edges of the rectangular region, elevation and soil electrical conductivity both changed in a gradual and correlated fashion. Those gradients contributed to the strength of the second principal component (Table 9–2) as a soil-based axis orthogonal to the crop-based axis (i.e., the first principal component). In the interior of the 6-ha region, crop and soil conditions changed dramatically in very short distances, showing sharp spatial discontinuities. The highest yielding zone (312 g m^{-2}) was immediately adjacent to the low-yielding spot identified by the farmer (zone mean: 212 g m^{-2}). The best crop health, as indicated by the aerial photography, occurred within meters of the worst crop health (first principal component score of −1.1 vs. 3.8), and the zone of highest soil electrical conductivity (81 mS m^{-1}) was within 20 m of the zone with the lowest conductivity (14 mS m^{-1}).

The farmer is considering the application of composted manure to the trouble spots in the field. Given transportation and application costs, the ability to isolate the most needy areas has a direct impact on the potential success of the application. Zones identified using the yield map data (Fig. 9–5d) illustrate some of the basic problems in yield monitoring. Along with the artifact created when the farmer stopped and restarted harvesting (the spot in the southeast sector of Fig. 9–5d), the zones around the problem area appear elongated in the north–south axis, a feature consistent with the way grain flow tends to become more uniform as it passes through the combine, due to the mixing that occurs at each step. The soil electrical conductivity measurements had the lowest spatial resolution, resulting in a stair-step appearance at their zone boundaries. The crop images provided the highest resolution data sources, and their application to spatial zonation resulted in patterns with the smoothest edges. The zone boundaries identified from crop imagery (Fig. 9–5c) appear to give the greatest precision in isolating the extent of the problem area. On the basis of our experience with this case and others in the Nebraska Soil Fertility Network, we recommend aerial photography as a means of capturing high-resolution information on crop performance, both for basic research on landscape spatial variability and for developing farmer recommendations. The map information in Fig. 9–5c provides the farmer with a basis for calculating the area over which he plans to apply compost (0.54 ha in the zone with the highest principal components score, and 0.26 in the next highest). Using his GPS unit, he can locate the zone boundaries and flag off the area where he plans to apply compost. The persons applying the compost need not have a GPS unit, and there is no evidence that a variable-rate application is warranted within the zone(s).

DISCUSSION

When conducting on-farm research, public map data may be the only source of information available for the fields offered. These data will not necessarily be current, as Fig. 9–6 illustrates. During the interval between when the picture for the digital orthophoto was acquired (1993) and the start of our on-farm experiment (1999), the farmer reshaped the crop landscape. Fortunately, the first image gives a basis for interpreting results obtained in this experiment. It is possible to identify field, pond, farmhouse, and waterway boundaries in the first image and test whether the treatment response (P fertilizer in this farmer's case) varies by position in the old landscape. The tendency of farmers to cultivate in the same direction (e.g., north–south in Fig. 9–6) and create management areas within fields that are long and narrow can generate errors in whole-field analyses if they are not considered during experimental design. Blocks should be aligned with previous land uses to help limit within-block variability. Tests for residual effects on treatment responses are also possible with proper blocking.

Historical aerial photography is a rich source of documentation for prior field uses. The original images used in Nebraska's first photo-based soil surveys are of excellent quality and suitable for GIS analysis. One problem with aerial photography is the time gaps between historical images. The interval between images is often too long to cover all the expected land use changes for our sites of interest. We also need additional management information to properly interpret the imagery. Integration of animal production and crop production is a particular concern. Anecdotal evidence shared by farmers and researchers (J. Schepers, personal communication, 2000) emphasizes the importance of knowing where animals were pastured and housed, where hay was harvested, and where manure was spread. Long-term effects of manure management can dominate maps of soil P availability. The success of soil sampling schemes for nutrient mapping depends on knowledge of that management history.

To assist farmers in documenting their management, we recommend farmers use a spatial record-keeping kit (Aschmann et al., 2003). The kit we have developed is a simple three-ring notebook containing a photocopy of an aerial image, protected in a plastic sleeve, for each field on the farm. Transparent plastic sheets are overlaid on the photos and markers are used to record notes and draw in boundaries for weed patches, flood damage, variety or planting date changes, etc. The kit requires no computer or GPS unit. By aligning the overlays with the aerial photos, GIS specialists will be able to digitize and georeference the records later. The kit is recommended for in-season record keeping and for capturing oral histories of field use. Recollections from elderly farmers, drawn by an assistant on the map overlays, would provide a unique data resource and provide us with opportunities for analyzing long-term agroecological dynamics across the time frame in which mechanized agriculture evolved.

One piece of new information technology could help automate spatial record keeping for farmers using digital monitors on their equipment. RAPID, Inc., a nonprofit organization, is currently assisting industry representatives to develop a public standard for communicating data on field operations. The standard is designed as an eXtensible Markup Language (XML) schema. Once published on the Internet in machine-readable form, the schema will provide computer programs with the definitions needed to interpret data stored in the XML format. Digital monitors mounted on sampling equipment, planters, fertilizer applicators, and combines will be able to store and transmit the technical information needed to document the de-

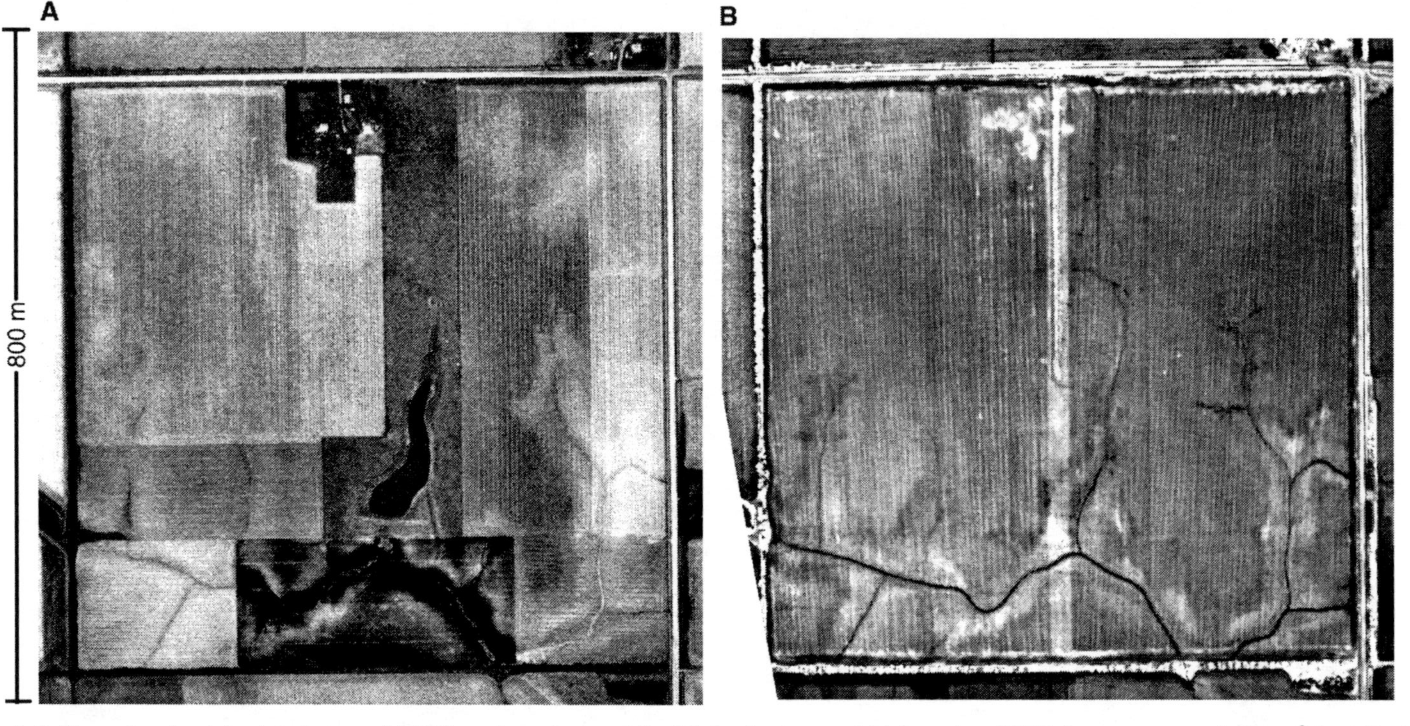

Fig. 9–6. Change in a farm's land use between (a) 1993, as shown in the public digital orthophoto, and (b) the spring of 1999 (image courtesy of mPower[3], Inc.). During that 6-yr interval, the farmhouse and pond were removed, a center-pivot irrigation system was installed, and the 65-ha area was converted to a single management unit.

tails of management actions, including times and georeferences, in a language that anyone will be able to interpret. Scientists conducting on-farm research will benefit from the labor savings afforded by automation and by the lowered risk of missing important management data. The standards could also benefit consultants who offer GIS analyses as part of their service. The consultants would be able to receive data sets via the Internet, interpret the data using the standards, conduct analyses, and send back reports and recommendations to the farmer. It is not clear yet what types of consultants will emerge to perform GIS analyses for farmers. Discussions at the 1998 Annual Meeting of the Nebraska Sustainable Agriculture Society and in subsequent Extension meetings indicated that people in rural communities with training in agroecology, many already using computers, may be well suited to perform GIS-based analyses for other farmers.

Holling (1992) used organism sizes, collected from a variety of ecosystems, to test his hypothesis that ecosystem dynamics are dominated by only a few processes expressed on the landscape in distinct spatial and temporal patterns. One counterpart in agroecology would be to analyze management unit sizes to see if "clumps" exist in the size distributions of the units across a variety of agroecozones. If clumps do exist, a next phase of research would be to identify why within-clump management unit sizes are well adapted to agriculture and why between-clump sizes are poorly adapted. This application to agriculture is not beyond the scope of the theory Holling and others are developing (Carpenter and Gunderson, 2001; Gunderson and Holling, 2001). Their vision is based on a fundamental cycle through which all adaptive systems evolve: the "systems renewal cycle" (Fig. 9–7). The cycle proceeds through four phases in which system resources are (i) exploited, through pioneers adapted to seeking and consuming resources; (ii) conserved, through strategies that maximize efficiency at the expense of redundancy and diversity; (iii) released, through breakdown; and (iv) reorganized. Figure 9–6 provides a pictorial backdrop against which we may hypothesize that the predominant movement in modern agriculture is within Holling's (1992) conservation phase, in which there is a loss of diversity (note the loss of the farm house, pond, and subfields) and an effort to increase and stabilize productivity (note the addition of the irrigation sys-

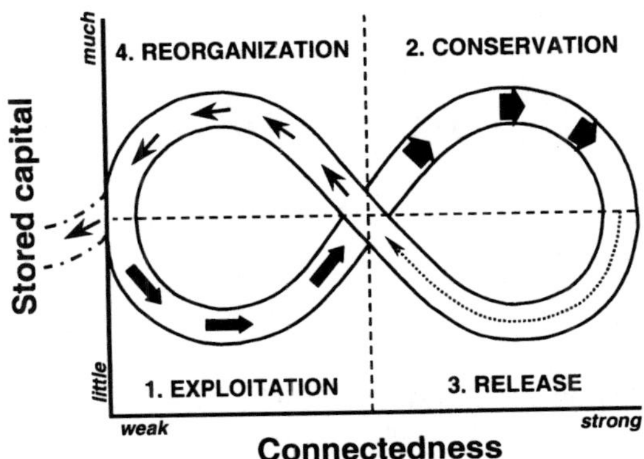

Fig. 9–7. Schematic diagram of the Holling's systems renewal cycle, redrawn from Holling (1992). The dimensions of the arrows reflect time spent in the phases, which is least during the release phase (e.g., fire or pest outbreak) and most during the conservation phase.

tem). If correct, the next phase would be dominated by fast and difficult losses to agriculture, paralleling the action of fire on grasslands and forests. The fundamental research we need to test this hypothesis requires use of GIS technology. We need to record, organize, and analyze the spatial patterns of resource use in agriculture, and develop theories that help us predict the long-term consequences of the patterns.

Farmer adoption of GIS technology and precision farming should not necessarily accelerate agriculture's movement through Holling's (1992) conservation phase. The technology could foster an increase in crop landscape diversity as farmers identify sections within fields that are never profitable under current management but may be adapted for other land uses, for example, including the incorporation of zones of grass or tree perennials. GIS analyses provide a rational basis for designing new, more diverse land uses. Compared with uniform whole-field management, precision farming can help fulfill the multiple objectives of increased farmer welfare, increased efficiency of input usage, and improved protection of natural resources in the short term and in the long term. On-farm research will likely be a part of precision farming as farmers adapt their management to the specific soil and weather conditions they face on their farm. To be successful in these uses of GIS technology and have an impact on agriculture, students of agroecology need to be trained in GIS use. GIS courses need to be offered as part of degree programs, vocational guidance needs to be provided for career opportunities in crop landscape analysis, and continuing education courses should be offered to teach state-of-the-art analytical methods. This investment in education and training will help ensure that the human component of GIS will be able to intelligently apply the power of the technologies to agroecology.

ACKNOWLEDGMENTS

The Nebraska Soil Fertility Network was initiated with Jim Peterson, Extension Educator, Washington County, Nebraska. Angela Mittan, Aaron Schepers, and Maribeth Milner helped process the data shown in the paper, and Marcus Tooze of GIS Workshop, Inc., assisted with editing. The research was funded in part by grants from the Nebraska Soybean Board and the UN-L Foundation.

STUDY QUESTIONS

1. How do Holling's ideas on ecosystem dynamics apply to agricultural lands?

2. Describe methods that can be used to describe "clumping" or "patches" in agricultural landscapes.

3. Why are past records and photos so critical in evaluating current landscape variability?

4. Devise a plan of action for evaluating an actual agricultural landscape that incorporates the ideas in this chapter.

REFERENCES

Aschmann, S.G., R.M. Caldwell, and L.B. Cutforth. 2003. Affordable opportunities for precision farming. USDA-NRCS Sustainable Agric. Tech. Note 190-1. Available at http://www.wcc.nrcs.usda.gov/watershed/pdffiles/Precision_Farming-Affordable_Opportunities.pdf (verified 28 July 2003). USDA-NRCS, Washington, DC.

Barnes, J.M., C.R. Trinidad, T.V. Orum, G.R. Felix, and M.R. Nelson. 1999. Landscape ecology as a new infrastructure for improved management of plant viruses and their insect vectors in agroecosystems. Ecosyst. Health 5:26–35.

Bellamy, J.A., and D. Lowes. 1999. Modelling change in state of complex ecological systems in space and time: An application to sustainable grazing management. Environ. Int. 25:701–712.

Caldiz, D.O., F.J. Gaspari, A.J. Haverkort, and P.C. Struik. 2001. Agro-ecological zoning and potential yield of single or double cropping of potato in Argentina. Agr. For. Meteorol. 109:311–320.

Caldwell, R.M. 2003. Analysis of cropping systems and use of decision support systems. J. Crop Production. 9:(in press).

Caldwell, R.M., and A.A.J. Fernandez. 1998. A generic model of hierarchy for systems analysis and simulation. Agric. Syst. 57:197–225.

Calera-Belmonte, A., J. Medrano-Gonzalez, A. Vela-Mayorga, and S. Castano-Fernandez. 1999. GIS tools applied to the sustainable management of water resources. Application to the aquifer system 08-29. Agric. Water Manage. 40:207–220.

Carpenter, S.R., and L.H. Gunderson. 2001. Coping with collapse: Ecological and social dynamics in ecosystem management. Bioscience 51:451–457.

Carsjens, G.J., and W. van der Knaap. 2002. Strategic land-use allocation: Dealing with spatial relationships and fragmentation of agriculture. Landscape Urban Plann. 58:171–179.

Chou, Y.H. 1997. Exploring spatial analysis in geographic information systems. OnWord Press, Santa Fe, NM.

Christakos, G. 2000. Modern spatiotemporal geostatistics. Oxford University Press, Oxford, UK.

Corwin, D.L., M.L.K. Carrillo, P.J. Vaughan, J.D. Rhoades, and D.G. Cone. 1999. Evaluation of a GIS-linked model of salt loading to groundwater. J. Environ. Qual. 28:471–480.

Creque, J.A., S.D. Bassett, and N.E. West. 1999. Viewpoint: delineating ecological sites. J. Range Manage. 52:546–549.

Crosetto, M., S. Tarantola, and A. Saltelli. 2000. Sensitivity and uncertainty analysis in spatial modelling based on GIS. Agric. Ecosyst. Environ. 81:71–79.

Environmental Systems Research Institute. 1994. Understanding GIS: The ARC/INFO Method, Version 7 for UNIX and OpenVMS. ESRI, Redlands, CA.

Farrow, A., and M. Winograd. 2001. Land use modelling at the regional scale: An input to rural sustainability indicators for Central America. Agric. Ecosyst. Environ. 85:249–268.

Franco, D., D. Franco, I. Mannino, and G. Zanetto. 2001. The role of agroforestry networks in landscape socioeconomic processes: The potential and limits of the contingent valuation method. Landscape Urban Plan. 55:239–256.

Frede, H.G., M. Bach, N. Fohrer, and L. Breuer. 2002. Interdisciplinary modeling and the significance of soil functions. J. Plant Nutr. Soil Sci. 165:460–467.

Gardi, C. 2001. Land use, agronomic management and water quality in a small Northern Italian watershed. Agric. Ecosyst. Environ. 87:1–12.

Gerstl, Z. 2000. An update on the Koc concept in regard to regional scale management. Crop Protect. 19:643–648.

Gunderson, L.H., and C.S. Holling (ed.) 2001. Panarchy: Understanding transformations in human and natural systems. Island Press, Washington, DC.

Hartkamp, A.D., J.W. White, and G. Hoogenboom. 1999. Interfacing geographic information systems with agronomic modeling: A review. Agron. J. 91:761–772.

Holling, C.S. 1992. Cross-scale morphology, geometry, and dynamics of ecosystems. Ecol. Monogr. 62:447–502.

Ines, A.V.M., A.D. Gupta, and R. Loof. 2002. Application of GIS and crop growth models in estimating water productivity. Agric. Water Manage. 54:205–225.

Jolliffe, I.T. 2002. Principal component analysis. Springer Verlag, New York.

Lagacherie, P., D.R. Cazemier, C.R. Martin, and T. Wassenaar. 2000. A spatial approach using imprecise soil data for modelling crop yields over vast areas. Agric. Ecosyst. Environ. 81:5–16.

Matthews, R.B., R. Wassman, J.W. Knox, and L.V. Buendia. 2000. Using a crop/soil simulation model and GIS technique to assess methane emissions from rice fields in Asia. IV. Upscaling to national levels. Nutr. Cycl. Agroecosyst. 58:201–217.

Microsoft. 2001. TerraServer [Online]. Available at http://terraserver.homeadvisor.msn.com/ (accessed 14 Jan. 2003, verified 16 July 2003). Microsoft, Seattle, WA.

Narumalani, S., Y. Zhou, and J.R. Jensen. 1997. Application of remote sensing and geographic information systems to the delineation and analysis of riparian buffer zones. Aquat. Bot. 58:393–409.

Nebraska Department of Natural Resources. 2001. Department of Natural Resources Nebraska map interactive [Online]. Available at http://www.nrc.state.ne.us/databank/interactive2.html (accessed 14 Jan. 2003, verified 16 July 2003).

Nizeyimana, E.L., G.W. Petersen, M.L. Imhoff, H.R. Sinclair, Jr., S.W. Waltman, D.S. Reed-Margetan, E.R. Levine, and J.M. Russo. 2001. Assessing the impact of land conversion to urban use on soils with different productivity levels in the USA. Soil Sci. Soc. Am. J. 65:391–402.

Priya, S., and R. Shibasaki. 2001. National spatial crop yield simulation using GIS-based crop production model. Ecol. Model. 126:113–129.

Robert, P.C. (ed.) 2002. Proc. of the 6th International Conference on Precision Agriculture and Other Precision Resources Management [CD-ROM]. Minneapolis, MN. 14–17 July 2002. ASA, CSSA, SSSA, Madison, WI.

SAS Institute. 1999. The SAS System for Windows. Release 8.00. SAS Inst., Cary, NC.

Steinhardt, U., and M. Volk. 2002. An investigation of water and matter balance on the meso-landscape scale: A hierarchical approach for landscape research. Landscape Ecol. 17:1–12.

Stockle, C.O. 1996. GIS and simulation technologies for assessing cropping systems management in dry environments. Am. J. Altern. Agric. 11:115–120.

Tucker, M.A., D.L. Thomas, D.D. Bosch, and G. Vellidis. 2000. GIS-based coupling of GLEAMS and REMM hydrology. I. Development and sensitivity. Trans. ASAE 43:1525–1534.

USEPA. 2000. Level III ecoregions of the continental United States (revision of Omernik, 1987). USEPA NHEERL, Corvallis, OR.

von Arx, G., A. Bosshard, and H. Dietz. 2002. Land-use intensity and border structures as determinants of vegetation diversity in an agricultural area. Bull. Geobotanical Inst. 68:3–15.

Walker, B., A. Kinziq, and J. Langridge. 1999. Plant attribute diversity, resilience, and ecosystem function: The nature and significance of dominant and minor species. Ecosystems 2:95–113.

Wu, J., and B.A. Babcock. 1999. Metamodeling potential nitrate water pollution in the Central United States. J. Environ. Qual. 28:1916–1928.

10 Serving Multiple Needs with Rural Landscapes and Agricultural Systems

CHARLES FRANCIS

University of Nebraska
Lincoln, Nebraska

LENNART SALOMONSSON

Swedish Agricultural University
Uppsala, Sweden

GEIR LIEBLEIN

Agricultural University of Norway
Ås, Norway

JUHA HELENIUS

University of Helsinki
Helsinki, Finland

One important contribution of agroecology is to provide a framework and an opportunity to focus on systems at the landscape level. The importance of recognizing multifunctional processes and products of rural landscapes has grown in recent years (Brandt et al., 2000). This is due in part to greater attention to systems dimensions of agriculture and food production, including emergent properties at scales larger than fields or farms. Researchers and practitioners attempt to weave new technologies into practical and economically viable production systems. Students seek relevance in their education, and meaning often comes from higher levels of scale. Farmers are seeking ways to design production strategies that preserve soil, water, and air quality, as well as biodiversity—numerous support programs are in place to encourage this direction. Broader interest and concern by consumers about the safety of food and the security of the food system for the long term bring additional attention to where and how food is produced, especially the rural environment (Torjusen et al., 2001). Other concerns, including global warming, limited supplies of critical resources such as fossil fuels, and the negative environmental impacts of many conventional agricultural practices, heighten people's awareness of the connectedness of the food system to the overall environment. Agroecology is emerging as an integrative discipline that will strengthen and support research and education on these complex issues in the agricultural production and food systems, as well as bring more attention to landscape-level analyses.

Agroecology, when defined as the ecology of food systems (Francis et al., 2003), is becoming a useful term that encompasses the complexity of resource use, the efficiencies of alternative production systems, and the social impacts of agriculture, including equity of distribution of benefits. When the scope of agroecology extends beyond the production field and the immediate impacts on water quality in nearby streams, we begin to evaluate whole food systems, including conversion of natural resources, efficiency of production, processing food items, marketing, and consumer issues. This allows an analysis of efficiencies through the entire system and provides tools to look at the global food chain in comparison with local food systems. Such an analysis includes the impacts of regulations and policies at all levels, as well as the potentials and applications of new technologies, and the overall environmental impact. Those who embrace agroecology with this broad definition must consider analysis of the impacts of consolidation and decisions by a handful of multinational food corporations that deal with commodities and global markets. It is useful to look at food systems at different levels in the hierarchy as well as through time. An example of a spatial hierarchy of agroecosystems, including social decisions at several levels, is shown in Fig. 10–1 (developed from a concept of Olson, 1999). We recognize that individual farmers and consumers make most decisions at the local (farm and community levels), and that only in the aggregate can these influence the global food system.

To study structure of a system at one level of the spatial hierarchy (e.g., field or farm or landscape), it is important to recognize the multiple characteristics that define agriculture at that level and how the factors interact to produce final food or feed products and economic returns. There are environmental and social impacts of agricultural systems at each level, including access to local resources, and community and ethical issues. More importantly, these factors interact across the spatial hierarchy and influence success or failure of production in complex ways. In addition to looking at different spatial levels, it is essential to bring in other dimensions such as time—the history and natural ecosystems that are sustainable in

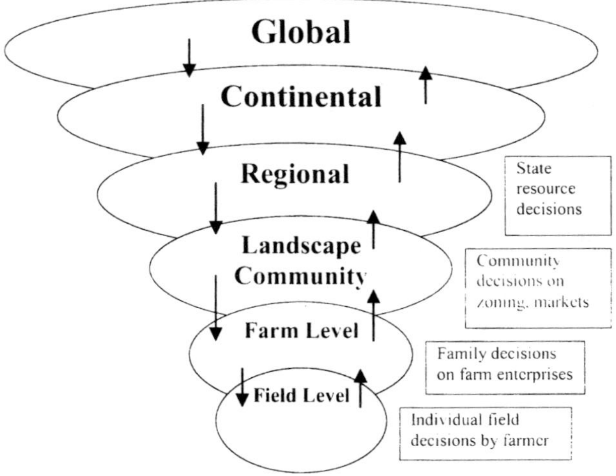

Fig. 10–1. Spatial hierarchy of scale with identification of social systems decisions at field, farm, landscape, and regional scales (inspired by Olson, 1999).

a given place; the current production and food systems characteristics; and the desirable future situation that will provide quality, safety, and equity in access to food into the future. These dimensions relate structure to function in the agroecosystem.

The performance of agriculture at the landscape level, often called landscape function, is highly influenced by decisions at the community, regional, and national levels. Political forces in each country and across the globe, including strong corporate interests that often transcend national political control and local benefits, influence markets, support programs, national priorities, and agricultural policy. More amenable to management, the key interactions within the landscape are conditioned by processes, mechanisms, and decisions at field and farm levels, as well as by the performance of individual crops and livestock species found across the landscape and the soil–climate complex that has evolved in each place. These in turn interact with components of the surrounding natural habitat. To understand how landscapes function, it is essential to understand mechanisms that operate at the farm and field levels (Olson and Francis, 1995). To appreciate the importance at the landscape level of broader influences, such as regional or global climate, international trade, or national agricultural policy, it is essential to look to higher levels in the spatial hierarchy to understand the context in which the landscape is embedded and the meaning that is attached to that landscape (Fig. 10–1).

Many examples of the growing attention in research to areas beyond the farm and immediate community were presented in the October 2000 conference, Multifunctional Landscapes: Interdisciplinary Approaches to Landscape Research and Management (Brandt et al., 2000), organized by University of Roskilde in Denmark. The conference attracted scientists and educators from a wide range of fields of study, and the participants struggled with their disciplinary dialects and biases in an attempt to deal with the broader issues that emerge at the landscape and regional levels. Many of the current methods and tools used by most agricultural scientists appear to be inappropriate at higher levels of scale, although we often persist in using the comfortable and well-accepted methodology of a given discipline.

We study these mechanisms and implications of how agroecosystems are structured and their major processes or functions in order to better understand how alternative agricultural and food systems work, and to use this information to better design systems for the future. This chapter explores structure and function at the landscape level. We then define the importance of mechanisms from lower levels and context and regulation forces provided by higher levels within a spatial hierarchy. We conclude with suggestions about the importance of designing sustainable landscapes. This is a greenprint for the future of agriculture, one based on the ecology of the food system and the impacts of this system on other spheres of human endeavor. The theme is expanded in the final chapter of this book (Chapter 12, Francis and Rickerl, 2004), which explores future themes and priorities for the study of agroecology and applications to food systems.

WHAT IS A LANDSCAPE?

The definition of a landscape is not precise, and in many ways a landscape is a dynamic phenomenon. The Webster definition is "an expanse of scenery that can be seen from a single viewpoint" (Steinmetz, 1993). Most definitions we encounter provide the perspective through the glasses worn by ecologists, other natural scientists, or perhaps agriculturists. Gliessman (1998) defined landscape ecology functionally as "the study of environmental factors and interactions at a scale that encompasses more than one ecosystem at a time." Forman and Godron (1986) presented five characteristics that are normally found and repeated in a landscape, obviously from a natural sciences perspective:

- Cluster of ecosystem types
- Interactions among the ecosystems in the cluster
- Climate and geomorphology
- Disturbance regimes in the clusters
- Relative abundance of different ecosystems within clusters

A landscape may be defined using these terms as "a heterogeneous land area composed of a cluster of interacting components that is repeated in a similar format throughout" (Olson, 1995). A landscape can be highly homogeneous as in the wheat (*Triticum aestivum* L.)–growing regions of western Nebraska or the Palouse in Washington, or in the vast maize (*Zea mays* L.)–soybean [*Glycine max* (L.) Merr.] regions of the U.S. Midwest. Landscapes can be highly heterogeneous, with crop and pasture mosaics embedded in natural forest areas, often where topography is more varied and promotes different types of enterprises. Often there is a mixture of land types and uses, these determined by individual farm owners and managers who choose the crops, animals, and rotations that best fit a given field or site. Olson (1995) pointed out that this is not random heterogeneity, as landscape components such as cultivated fields, linear elements such as roads and fencelines, trees and shrubs growing along water courses, forested uplands, and farmsteads are repeated in similar spatial patterns. What is important is how the structure of these managed rural landscapes influences their function, both the immediate economic returns to owners and the ecosystem services that are provided for the wider society (Daily, 1997).

This same geographic landscape could be described differently if viewed through a set of "sociocultural glasses." The cultural landscape would have areas of sparse human habitation, depending on farm size and how many operators live on the land, as well as rural dwellings of nonfarmers. Communities in the landscape are denser aggregations of people, along with associated infrastructures of business, manufacturing, social services, and other urban activities. This social landscape overlies the natural and farmed landscape described above, and the interaction of the two results in agroecosystems and associated human population centers important to the function of nearby farmland.

Without a precise definition of size, it may be difficult to conceptualize landscapes and discuss their characteristics as well as their functions in more concrete terms. Landscapes could be considered areas from 10 to 1000 km^2 (e.g., Risser, 1987, cited by Olson, 1995). In U.S. terms this is about three to three hundred sections, or three to three hundred square miles, a range in size from one large farm to one moderately small county. Some variables (e.g., water quality in a stream coursing past production fields and then used for livestock water) could be studied within a farm or across contiguous farms, while others (e.g., community dynamics and emergent properties from multiple farms) must encompass study at the higher end of this scale. Calling a landscape an area "larger than a breadbox" but smaller than a watershed does not add precision, but it does allow us to focus on an area that functions in an integrative way and has some of the characteristics described above (Olson, 1995).

Time is another critical dimension in landscape function, and Olson (1995) pointed out that fields change from year to year, farms shift focus to crops that are more economically viable, forests are cut or thinned, and linear elements are added to or removed from the landscape. These changes are due to relative crop and lumber prices, availability of new crops or technologies, and/or government support programs, among other factors. Changes at the farm level happen more quickly than those at higher levels in the scale, and crop acreage at a level such as county will shift less drastically in the short term unless there is a major change in policy or markets. When we observe a landscape—recall the above Webster definition—this is a snapshot at one point in time. To enhance this vision, we must learn the history of this place through written, visual, or oral records, and observe carefully the effects of prior management decisions on current functions in the landscape. Similarly, communities change in size and function, as an economy of each area grows or contracts. Age structure of the population changes, and with that the nature and function of community. For example, the median age of farmers in the U.S. Midwest is about 58 yr, and many small rural towns likewise have an aging and shrinking population. This strongly influences the nature of services needed, buying habits, and connections to the rural landscape.

In the European discourse about landscape management, an interdisciplinary approach is highlighted, encompassing natural, social, and humanistic sciences. Giorgis (1995) proposes four fundamental principles to be considered for landscape quality:

- Respect for life and preservation of landscape diversity
- Preservation of biological diversity
- Development of solidarity
- Respect for regional identity and the right to enjoy beauty

These encompass some of the ethical and aesthetic goals of human residents in the landscape and clearly represent a view through a set of socioeconomic glasses.

MULTIDIMENSIONAL CHARACTERISTICS OF RESOURCES

Natural resources as well as those generated by society can be described as they occur and as they impact agriculture and food systems at any given level in the spatial hierarchy. To classify resources according to any scheme is obviously an arbitrary human construct that is intended to help us better understand the world and its agricultural systems. We find this is useful for exploring the intricacies of structure, function, and factors that contribute to system interactions and to inform the design of more efficient and sustainable systems for the future.

Resources can be discussed and evaluated as those that occur in the natural system and those in the social system, as shown in Fig. 10–2 for any one level of

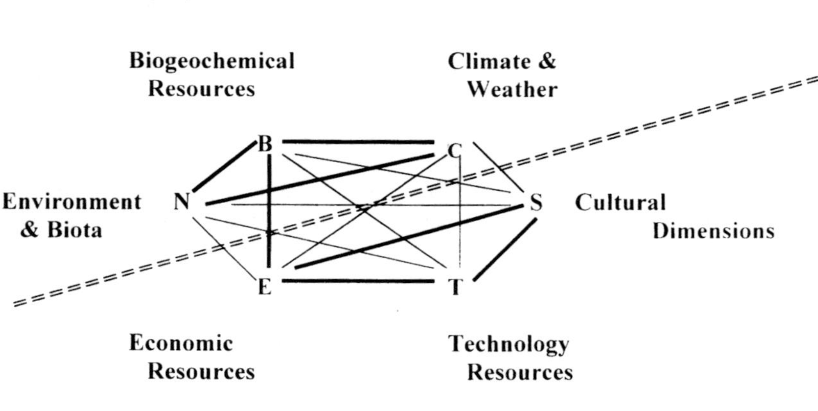

Fig. 10–2. Multidimensional characteristics of resources in an agroecosystem and their interactions that lead to multifunctionality, shown for a single level in the spatial or geographical hierarchy of scale.

the spatial hierarchy. In the upper and left are natural systems components such as geochemical resources (i.e., the geological complex that forms the land structure and chemistry), climate and weather, and the biology of the natural and farming environments that interact with the first two. In the lower and right are social systems, including economic resources (i.e., power to allocate other human efforts), technologies (i.e., both the physical structures and the ideas and aggregated information that make them), and cultural dimensions (i.e., human information aggregated and tested for a long time span in a defined context) that are usually unique to community, region, or nation. Shown also are major interactions among these categories (bold lines) and other interactions that contribute to the complexity of the system at any level in a spatial scale.

Interactions within the Landscape: U.S. Examples

Important interactions are first described at the landscape level as an illustration, but there are also similar interactions of various magnitudes at all levels of scale (see Fig. 10–2). Specific examples include characteristics of major cropping systems in the Midwest and elsewhere. Primary interactions are found between climate and weather and the geochemical or soil resources in any landscape. Soils are formed from parent materials over centuries by interactions of geological and climate processes, such as mountain formation and degradation, volcanic eruptions, glacial ice during ice ages, as well as the prevalent rainfall amount and pattern and attenuated by the temperature and winds that occur seasonally in each place (see Chapter 3, Francis, 2004). Loess soils along the Missouri River in Nebraska and Iowa were deposited over millennia by wind-blown soil particles originating farther west. The interactions among these soil and climate factors result in unique local environments, biotic complexes, and natural ecosystems. The tallgrass, midgrass, and shortgrass prairies that appear on a transect across Nebraska are the result of rainfall, temperature, and the soils that dominate in each part of the state.

In the social systems, there is a strong linkage between economic resources in a landscape and the access to technology by people living there. Culture also influences the adoption of technology, as shown by differences in use of new innovations by Amish groups in the Midwest compared with large corporate farms, often in the same landscapes. There is also an important link between economic level and culture, as illustrated by a comparison of the industrial-model farm that raises commodity crops across several thousand acres with the small family farm growing vegetables for local, direct sale from a few acres. All these systems are present and interact in the rural landscape, even though there are great differences in the areas of land occupied and influenced by different farmers and managers, and in the amount of production per farm. Most often, the benefits and the impacts on the economy of a landscape are not uniformly distributed.

One of the primary links is between soil productivity and economic resources, as shown by the relative wealth of farms in areas with deep, fertile soils such as central Illinois compared with those with highly weathered soils on hills in southern Missouri or those made saline by decades of irrigation and high evaporation such as the Imperial Valley of California. Soil quality may be influenced by availability of appropriate technologies for tillage or remediation of nutrient deficiencies by fertilizer or animal manure in mixed farming systems. In a related example, many areas of the Sahel are low in productivity due in part to lack of P, while cultural factors such as high status attached to size of family cattle herds can lead to overgrazing and further deterioration of soil productivity.

Climate in a given location determines economic success. For example, well distributed rainfall during the summer growing season in the eastern Corn Belt of the U.S. leads to relatively consistent crop yields and financial benefits to farmers. Unpredictability of weather can impact decisions on technology, such as purchase of larger tractors and planter units in Nebraska to accomplish timely planting when conditions are near optimum. Low-input organic systems appropriate to a given climate niche are used by some small farmers to reduce risk and increase value of products. Climate shapes human behavior in agriculture by determining length of growing season, available days for field work, and thus the types of complementary activities (on- or off-farm jobs) families can accomplish while doing a good job with farming. These are examples of interactions between pairs of factors shown in Fig. 10–2. It is obvious that many outcomes in farming are the result of more complex interactions, for example soil characteristics with timing of rainfall with available technology and labor for planting. The figure only indicates some of the most simple connections between factors at the landscape level.

Interactions within the Landscape: Nordic Examples

The most fertile soils in the Nordic countries, the clay soils, were formed by the force of the glacial ice during the last ice age. The moving force from the glacial ice crushed stones down to clay size particles, and the weight pressure of the ice pressed the land area down, setting the stage for the following land lift, when the clay sediments were lifted from the lake bottoms to become marsh lands and later the flat, grassland areas that we are now farming.

Another prime determinant of land use at the landscape level in Nordic countries is the location of major cities on or near prime farmland and also near a harbor. Major cities such as Oslo, Trondheim, Gothenburg, Malmö, Copenhagen, and Helsinki all expanded due to excellent harbor potentials for trade at the expense of flat and fertile land for agriculture along the coasts. This is an interaction among soil resources, economics, and sociocultural factors. Favorable climate and fertile soils in south-central Norway and the proximity of this land to modest-sized communities provide markets for a wide range of vegetables and fruit crops, and thus interactions between natural factors and the social system at the landscape level.

Proximity of beautiful natural areas and farmland to the population centers such as Copenhagen, Roskilde, and Aarhus in Denmark make these highly susceptible to residential development, as people migrate from the countryside to larger cities in search of urban jobs and greater cultural opportunities. Development of excellent transportation services, especially dependable train schedules, allows people to work in a population center and live up to an hour's travel away (100 km) in a rural community. Although this maintains a population in the smaller rural communities, the nature of business and social activity changes in that economic landscape since many purchases are made in the larger nearby city, reducing commerce and economic viability in the bedroom communities. Increased purchase of private automobiles by most families in the Nordic region during the past two decades has further eroded small town economies as people travel quickly to shopping malls in nearby cities. Small, locally owned shops in Ås, Norway often close or change management as a result of pressures from three such regional malls within 15 km travel by car. Economic factors interact with technology in the changing culture at the landscape level.

SPATIAL HIERARCHY AND LINKAGE OF FUNCTIONS ACROSS LEVELS

There are obvious connections between agroecosystems and across levels in a spatial hierarchy within a given function or characteristic of the system, as illustrated in Fig. 10–3. The biogeochemical resources may be represented by two or more soil types within a field, although with relatively uniform topography there may be a single type in some fields. The aggregate of different soil types in several to many fields make up the more complex array of soils on a farm in the U.S. Midwest: some with deep and fertile soils of high productivity potential in lower areas near streams, other fields with shallow topsoil and less water holding capacity on hillsides, and relatively fertile areas on the tops of hills where less recent erosion has occurred. These are examples of interactions between field and farm levels.

When a group of farms is viewed as part of a landscape, we find these same soils repeated in similar ways in similar topographic situations, with soil type determined by slope, depth of topsoil, drainage, and position in the landscape. More soil types with a wider array of potential for crop and pasture growth occur at the regional level, as this encompasses numerous landscapes with more variable rain-

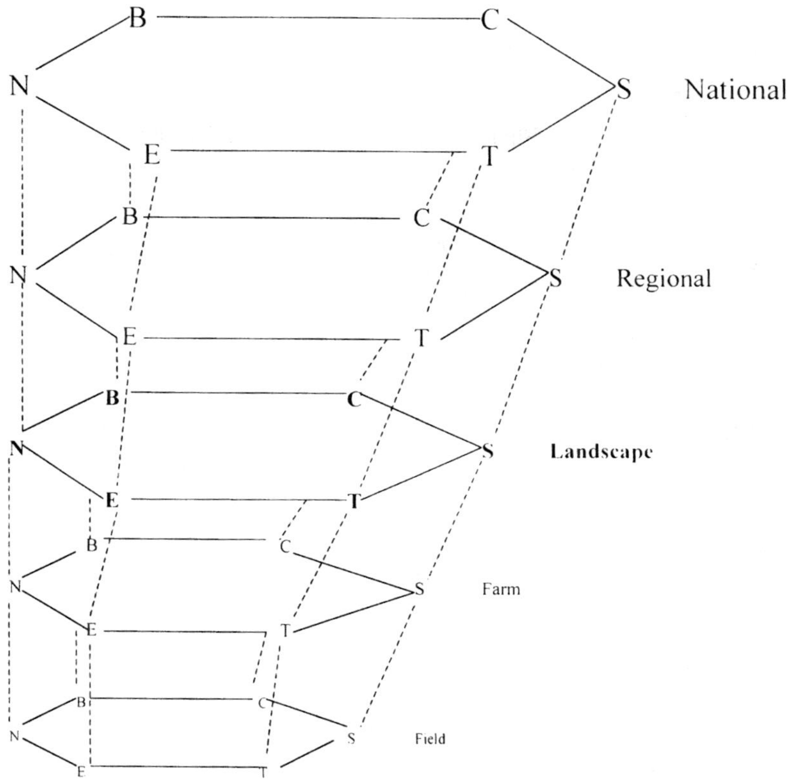

Fig. 10–3. Spatial hierarchy of characteristics of food systems and agroecosystems, not shown to scale, and aggregations across the spatial scale.

fall, proximity to oceans, and a wider range in physical relief. At the national level, a wide range of soil types is found, depending on the size and location of the country.

Similar aggregation can be described for climate, especially for the unique weather patterns in each landscape and even each farm that can be quite spatially variable within seasons and from year to year. Environments and biotic elements are somewhat uniform at the field level, especially with production of monoculture commodity crops as in the U.S. Midwest. There may be more diversity at the farm level, if multiple enterprises including crops and livestock are found in a mixed farming situation such as northeast Nebraska and southeast South Dakota. Progressively more component diversity is found as we view the landscape, the region, and the national levels. Thus, the characteristics within a given category are nested within the higher levels of the hierarchy.

Even more intriguing are the connections of elements of one factor at one level with those of another factor at another level. In Fig. 10–4 there are several examples. Oil reserves at the national level (a biogeochemical factor) can easily influence the price of diesel fuel on farms, and thus the choices of technology at all levels in the hierarchy. The entire loss of fossil fuel supply caused Cuba to move toward

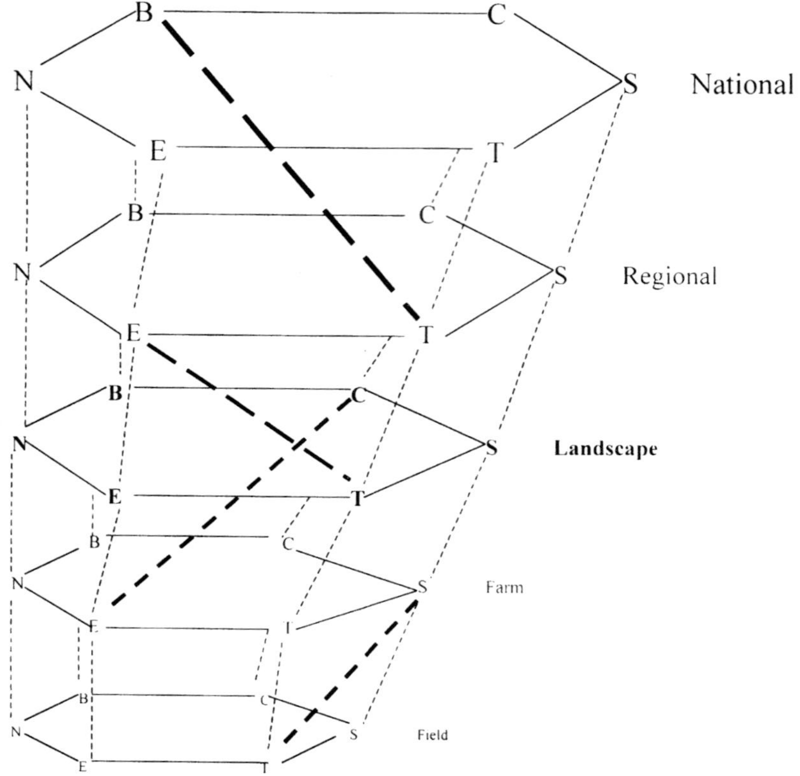

Fig. 10–4. Examples of interactions across the spatial scales between different factors that are important at the different levels.

animal traction, reliance on small-scale food crop cultivation, and more local food systems in a few years in the early 1990s (Deere, 1997). Economic success in farming at the regional level can strongly influence the acceptance or rejection of new technologies at the landscape level. When the farm economy is strong, there are ample supplies of new equipment at local dealerships, and this further spurs other business in that community. When the economics of farming are fragile, there is a greater supply of used equipment, and greater sales of parts to keep existing tractors and implements in service.

Climate and weather at the landscape level can strongly influence cropping success on each farm in that landscape, and thus the financial conditions on each farm in each season. Social factors such as labor available from the farm family (more labor when children grow to an age for driving equipment, or less labor when they leave for college) may be strong determinants in what technology will be used or what crop enterprises are appropriate for a given field. There is a multiplicity of interactions across levels in the hierarchy, and these examples illustrate the types of interactions that can influence decisions on the farm and in the landscape or community.

Interactions across Levels: U.S. Examples

In the U.S. Midwest, dairy farming was once practiced in small economic units integrated with crops on many farms. One often hears that when all the children left the farm for college or other jobs (a change in labor, a social factor at farm level), it became necessary and logical to change to less labor-intensive beef cattle or to abandon livestock entirely (enterprise and technology change at the field level). When an elevator closes in the nearest town (a community or landscape level change), this may require individual farmers to change enterprises (field level) or restructure the entire operation (farm level) to accommodate crops or livestock that can still be marketed efficiently.

Economic or regulatory decisions, such as limits on the allowable nitrate in groundwater or physical distance requirements for locating a new well (regional political decisions), may strongly influence cropping systems and practices across the landscapes within that region. Farmers may shift to growing less N-demanding crops, such as from corn production to corn–soybean rotation or pasture, or to crops that require less irrigation, from corn to grain sorghum [*Sorghum bicolor* (L.) Moench]. Conversely, periods of greater or more consistent rainfall at the regional level for several years such as in the past decade in eastern Nebraska have stimulated farmers to replace grain sorghum with corn. National economic decisions on a farm support program to target several key commodity crops can strongly influence the areas in those crops at regional, landscape, and farm levels. The Conservation Reserve Program (CRP) has become popular as farmers age and children leave the farm, an interaction of policy with family dynamics over time. These are current examples of interactions across levels in U.S. agriculture.

Interactions across Levels: Nordic Examples

The regionalization of milk production in Norway is a prime example of a political decision that impacted structure and functions of farms. In the late 1950s and early 1960s the Norwegian government launched an agricultural policy that essentially enforced a regional specialization of farm production. Through economic incentives grain production was increased in the best agricultural areas, mainly in the southwest part of the country. Milk production and other animal production were moved to the valley and mountain farms of the country (Lieblein et al., 2001). The decision was effectively enforced by providing quotas and subsidies for production in those designated areas and none in the rest of the country, resulting in a milk price more than twice as high for those dairies with quotas. Milk and other cattle production was greatly reduced in the grain cropping areas, leaving a large and valuable infrastructure in barns and other facilities not used. Although farmers were compensated for their losses and helped to convert to grain production, the change caused a large restructuring and social upsetting of the entire farm sector. Another impact was the spatial separation of animal manure generation from the large grain crop areas. In addition, regions specialized in milk production import an average of 40% of their feed from the grain crop areas of the country. As a result, the distance between the producer and the consumer has increased, and the recycling of animal manure has been greatly hindered. At this same time following World War II, the large and financially powerful Norsk Hydro fertilizer company began to grow, and this commercial venture provided the N and other nutrients to restore soil fertility in areas where manure was not available.

In Norway, the government has decided that the acreage in organic production should be 10% of the total acreage by the year 2009 (Det Konglige Landbruksdepartement, 1999). This ministry of agriculture has recognized that as a result of the regional specialization, the growth of organic farming is not sufficient in the typical field crop areas because of the low presence of cattle. Therefore new quotas for milk production have been given to farms, but only to those farms that would start up organic milk production.

Another example of interactions across hierarchical levels comes from the European Union (EU) subsidy structure and its influence on enterprises and practices at the farm and field levels in Finland, Denmark, and Sweden. There are many positive incentives for conservation practices, limits to fertilizer application and manure spreading, and conversion to ecological methods (Lohr and Salomonsson, 1999). These are intentional economic decisions that impact both technology use and soil fertility at the farm and field levels. With close to one-half of the EU budget going to the Common Agricultural Program (CAP) payments, there is concern about the sustainability of this intervention and the high costs of administration. Swedish farmers today say their most important consultant on the farm is the one who understands and helps them fill out forms for support programs, a situation not unlike that in the USA. As an example of landscape effects in Sweden, the growth of flax (*Linum usitatissimum* L.) for oil was strongly supported by EU subsidies in 1999 but not in 2000, raising the national production of flax seed from about 6000 t (1998) to 33 000 t (1999), and then down to 8000 t (2000). This was recognized by the public in several comments in newspapers and media about the nice blue "new" crop, that came and then disappeared so suddenly.

In Lithuania, a national policy of open trade is moving their markets and agriculture closer to current regulations and standards in the EU. The goal is to become a full member of that regional organization in the near future. While this national level goal of regional integration is perceived by some as desirable in the long term, the present impact on local agricultural profitability is obvious. About 20% of the nation's farmland is currently idle due to lack of economic incentives, and the negative effects on a newly developing free market economy as well as on food security are hard to assess. Similar situations exist in Latvia and Estonia. Policy decisions at the national level affect farm and field level decisions by farmers and the ecosystem functions on these lands.

Finally, the prevailing opinion about food safety, a cultural factor at community, regional, and national levels, is influencing the organic food demand and prices in northern European countries. Recent problems with bovine spongiform encephalopathy and foot and mouth diseases in livestock have spurred the markets for locally produced, organic meats and other products in the U.K. and Germany, among other countries. This impacts farm-level decisions on whether or not to convert to organic production or whether to expand to meet these new demands. In Norway, in contrast, there is prevailing opinion that the conventional food supply is adequately safe, few pesticides are used, and a long winter make these problems less important than in countries further south. Although the demand for organic food is growing in Norway, it is not nearly as easy to market products at a premium as it is in other countries in the region. These are public opinion (social) factors at the regional and national levels that influence the economics at farm level. In summary, there are many factors that interact across levels in the spatial hierarchy, and these are often complex and change with time.

MULTIFUNCTIONALITY OF THE RURAL LANDSCAPE

These interactions among factors at any level in the hierarchy of scale and across levels are operating within a complex rural landscape that performs many functions for farm families as well as for society. Growing awareness of this breadth of functions is leading to a new appreciation of the importance of ecosystem services that derive from the rural landscape (Björkland et al., 1999; Drake, 1999). These functions have been recognized and rewarded for several decades in northern Europe, while this is a relatively new concept in North America (Daily, 1997).

Production of food, fiber, fuel, and other raw materials has been recognized as the primary function of agricultural landscapes. Roles of the natural landscape in receiving and storing water from rainfall and snowfall, preventing floods, cleaning the air by plant photosynthesis (converting CO_2 to O_2), and providing a haven for biodiversity are becoming better recognized and valued by society. We now recognize that virtually all landscapes worldwide are influenced by people in some ways. Thus, we have the obligation to preserve these landscapes and their vital functions (Baskin, 1997).

Since many of the best lands and those nearest human communities are farmed with higher levels of intensity and many hectares are converted to nonagricultural uses each year, we need to understand what functions in addition to production these lands provide to society. Next it is important to design farming systems that can enhance rather than reduce the effectiveness of these functions, including crop rotations, crop–animal integrated systems, and more diverse enterprise mixes on each farm. To promote this process it is essential to provide financial incentives to private land owners, in combination with regulations, to assure concern and compliance. Functions of rural landscapes where agroecosystems predominate include water filtration and storage, carbon capture and storing by annual crops and perennial plants (carbon sequestration), and mitigation of extremes in climate such as preventing floods and reducing wind speed with vegetative barriers. These kinds of ecosystem services are a collective utility to society that has no current market. The main work to produce these utilities comes from nature and is difficult to value with money. The loss of income and extra work for the farmer who opts for alternative land use to promote these services needs to be valued in monetary terms and paid for in some way by society.

Federal water quality standards for drinking water consumption in the USA (e.g., 10 parts NO_3–N per million) are one example of a regulation and a goal that all cities must achieve, and which is highly recommended for people with private wells. Communities can often receive federal and state assistance in upgrading a water system to meet standards, such as digging deeper wells, installing special equipment, or locating a new water source. Less often, we try to trace the source of the NO_3 and solve the problem upstream in surface waters or in underground aquifers.

SEARCH FOR MECHANISMS AT THE FARM AND FIELD LEVELS

Although most educators and researchers in agroecology have expanded their focus to the systems interactions and implications in agriculture, it is essential that we continue to search for understanding of mechanisms at the field and farm levels, as well as recognize the importance of regulating forces from higher levels. As shown in Fig. 10–1, the need to understand how systems function is met by study of mechanisms at a smaller or lower level of scale. To understand success or problems on the farm, either in production or economics, we need to look in detail at the individual fields or enterprises that make up that farmer's system. To understand the performance of a crop in the field, it often helps to examine the insect damage or the yield components of individual plants. The growth and development of plants depends on nutrient uptake and soil biota. This work needs support and focus. A cursory visit to poster sessions or presentations of agricultural professionals in national and international meetings will convince the observer that this activity is alive and well. Relative to systems research, component investigation is well supported and remains the predominant activity in our research profession.

Two issues in research on mechanisms are important to this discussion. The first is that greater relevance and potential application can be achieved in research on system components if these are clearly identified with elements that improve productivity or economic return, reduce negative environmental impact, and improve the distribution of benefits of the research. Too often we assume that increasing yields will automatically cause good things to happen: higher income, greater efficiency, more food for a hungry world. Most of us trained and practiced in a research environment where higher yields were intrinsic in our goals. We often failed to look beyond that goal to evaluate the potential impacts of the research. It is highly desirable to conduct component and mechanistic research within a systems framework. The other issue is the opportunity cost of research emphasis on narrow component technologies, especially on the subset of innovations that have potential for rapid commercialization. Focus on herbicide-resistant crops and other genetically engineered traits in new varieties has obvious benefit to the seed industry, but questionable value to the farmer. This research emphasis precludes putting the same resources into systems research for reducing weed pressures to below economic levels and even breeding for other traits, including higher yields. Research on global positioning equipment and site specific management has the stated goal of reducing input costs, while in fact these new technologies have a high cost and are not scale neutral in application. They provide another technological fix that pushes for larger farm size and less common sense management by farmers who know their fields well because of frequent visits and decisions based on long-time experience and observations. Component research needs to be relevant to both immediate and long-term challenges at the field, farm, and community levels of agriculture and the food system. Equally important is the continued quest to encourage and support systems research at all levels of the spatial hierarchy.

IMPLICATIONS AT THE REGIONAL AND NATIONAL LEVELS

The search for meaning or context in agricultural landscapes, as well as the major regulatory forces, and the potential to recognize and reward multiple functions are found at higher levels in the spatial hierarchy (Fig. 10–1; also Doherty et al., 2000). One contribution of the perspectives in agroecology to decisions at the regional and national levels is the multidimensional analysis of problems and the search for holistic and long-term solutions. This is in direct contrast to the linear cause and effect analysis and singular solutions often used to meet food systems and other challenges. To solve a water quality problem of pesticide residue or NO_3 contamination, strict limits may be placed on these inputs across a wide geographic area without considering the variation in soil types, insect populations, or cropping patterns that exist in the area. Regulation may spur interest in seeking solutions through research, but frequently this is focused on an alternative product for insect control or a new formulation for chemical fertilization. Because of our dedication to the technological quick fix, few researchers step back and look at the whole system and how it might be modified to alleviate the problems. A broad view of problems at the higher levels in a spatial hierarchy may reveal a wide range of options for how to modify agricultural systems to make them more efficient and profitable for the longer term.

One useful example is the contrast between a global food chain and a local food system (see Fig. 10–5). In the current expanding global markets and food import–export volume there is emphasis on purchased technology and inputs, and production is often separated from processing by large distances. Likewise, the marketing business is worldwide and much food travels a long distance to reach the consumer. The average food molecule in the U.S. system travels 2400 km (1400 mi) from point of production to the consumer. This requires an extraordinary investment in transportation infrastructure and fossil fuels to keep the system viable. Also, there is little potential for materials cycling at any step in the process. Waste from processing and marketing is more easily discarded, at some cost, to landfills rather than recycled into the production process. From the consumer, it is virtually impossible for any materials to reach the farm, unless there is a local program to apply sewage sludge to farmland. More often this also goes to the landfill for economic or regulatory reasons.

A serious and comprehensive analysis of the potential materials use and cycling in the food system could lead to greater efficiencies, sustainability, and security in the food system. Closer physical proximity of all activities allows obvious efficiencies such as direct sale from producer to consumer or processor to consumer. When these systems activities are spatially and temporally close, and even related to each other through management or ownership, a design for materials cycling is possible. By-products or compost from processing or unsold products in the market can be returned to the farm under some mutually favorable financial arrangement. Some wastes can be used directly as animal feed. Most organic waste can go back into the nutrient cycles, including that from consumer food preparation and from human waste. These cycles can be promoted by local and regional regulations that allow cycling with proper health safeguards, and that provide incentives for these steps to occur. The argument that we could not have bananas (*Musa*

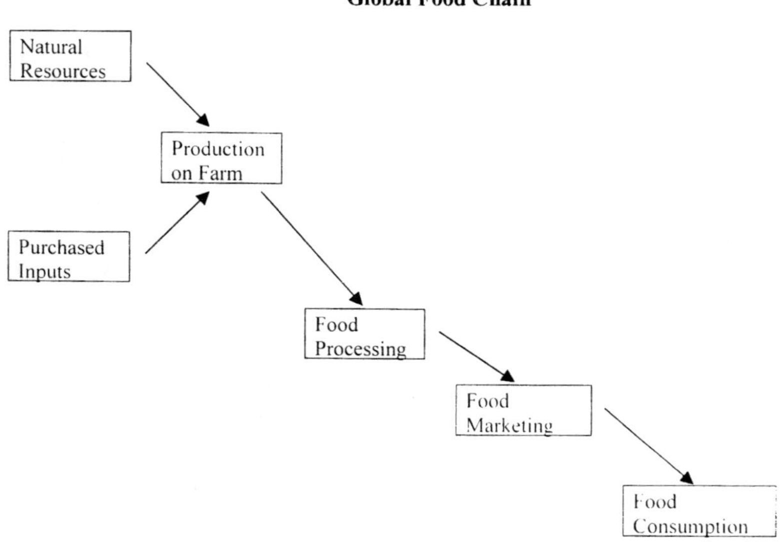

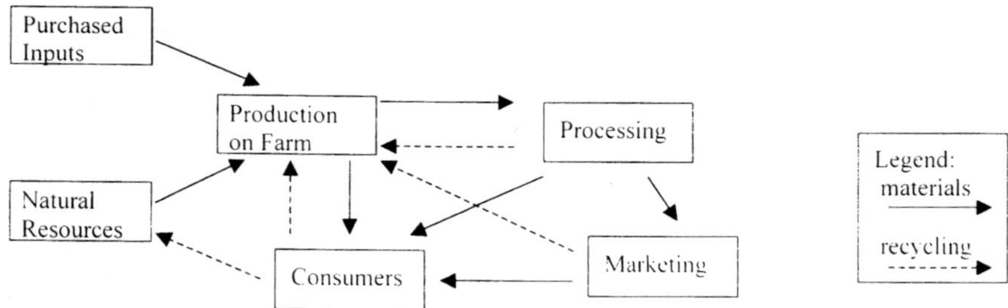

Fig. 10–5. Contrast between a global food chain and a local food system or cycle.

× *paradisiaca* L. var. *paradisiaca*) or coffee (*Cofea arabica* L.) in Nebraska in such a local system is really a "straw man." The objective is to substitute imports whenever possible with locally grown foods and to make this both financially and psychologically attractive, and not to exclude all imported food from any local system.

Such changes require creative thinking and planning at all levels, and not a slight modification of business as usual. According to Albert Einstein, "It is not possible to solve today's problems with the same ideas and people who created them."

CONCLUSIONS: DESIGNING MULTIFUNCTIONAL LANDSCAPES

There is a high level of recognition of the importance of multifunctional landscapes and the need for regulation and legislation to encourage this type of landscape use in northern Europe. It is intriguing to consider that the same basic human genetic stock in North America has moved in a somewhat different direction, one that initiated and developed a high-technology agriculture based on virtually unlimited land and resources at least in the initial stages. The different directions taken by many of the same people could be explained by the environmental determinism described by Diamond (1997). We must recognize that Europeans have been in the food production business for more than six millennia, while U.S. industrial agriculture has developed in less than two centuries. We all need to gain perspective on how to manage scarce resources.

One method of increasing income to farmers as well as solving negative environmental impacts from agriculture is the move toward organic farming (often called ecological agriculture in northern Europe). Norway has set a national priority of 10% organic food within 10 yr; Sweden has a national goal of 20% within 5 yr. Denmark has enough local production to meet the current demand of about 20% for organic milk and meat products, perhaps near the highest level in Europe today. Denmark has also dedicated more than DK50 million ($8.5 million) and Sweden more than SEK 30 million ($3 million) annually to research on organic farming, and there are more than 100 scientists involved in this research. Switzerland and Austria have national policies to encourage organic food production. Such prioritization at the national level attracts many of the best and brightest to this research area, just as genetic engineering attracts young scientists in the USA. Clearly, national public priorities and research emphases can influence and broaden the research agenda, while that controlled by corporate interests will promote those products or systems that can be patented and provide profits, and thus narrow the agenda.

Multifunctional landscapes in rural areas are an obvious current reality, but we need to explore creative and effective methods to encourage those who own the land and manage agroecosystems to maintain biodiversity and otherwise preserve ecological functions that serve society. Payments for conservation plantings can encourage perennial species and woody buffers, and some practices can promote better preservation of crop residues on the soil surface to reduce erosion. Regulations that promote local food systems and materials cycles are useful. Promoting the awareness of community economic advantages to local systems can facilitate the process. Land owners can be encouraged to diversify with enterprises beyond food production, such as rural recreation, hunting and fishing rights, farm weekends, educational activities on the farm, and public awareness campaigns. Society should pay for those services that provide for the common good, such as flood protection, carbon capture, and clean air and water. Rural open areas should be preserved for their ecological functions, historical relevance, and aesthetic values to society. Most European countries have strict rules as well as government support to preserve for posterity the rural cultural landscape.

The design and implementation of sustainable and multifunctional landscapes requires input from a range of disciplines beyond those associated with production agriculture. Natural resource specialists who deal with water, soil conservation, and wildlife habitat are essential to the team. Rural sociologists and community planners are important to understanding human motivations and the potentials for effective design alternatives for human communities. Legal experts can address the complexity of regulations and individual rights that govern building and resource use in the rural landscape. When these people and groups work together toward common goals, their results will be best coordinated and most effective. Combining natural science and social science methods is essential. This strategy is strongly supported by a white paper developed at a June 2000 workshop supported by the U.S. National Science Foundation (Kinzig et al., 2000).

Finally, the need to look at the landscape as a complex and interactive living system is obvious. It is more than a series of fields and farms connected by tree lines, roads, and streams. The landscape is dynamic and sensitive to human management. We need a proper mix of better systems education, adequate financial incentives within the current economic system, and an awareness that we all have a vested interest in a healthy, multifunctional landscape. To not address these challenges and seek innovative solutions would be to foreclose some options for future generations. We would do well to follow the traditional Native American wisdom and look at the impacts of our decisions today on the landscape with respect to how this will affect people in each place seven generations in the future. Short-term landscape planning will not achieve our long-term goals.

STUDY QUESTIONS

1. List and discuss the ecosystem services that are most important in your landscape, region, or state, and describe how these are important to the human and broader biotic community.

2. How can the multiple functions of agricultural landscapes be recognized and rewarded by society? What systems of monetary or tax benefits could be used to encourage private rural land owners to preserve and enhance ecosystems services?

3. Why is it useful to study agroecosystem activities and services within a given level in the spatial hierarchy? How do these things change through time?

4. Why is it essential to study systems structures and processes across levels in the spatial hierarchy? How does this type of study help us understand system resource use and productivity?

5. Discuss the unique emergent properties that characterize the agroecological landscape, and why these are more difficult to observe or study at lower levels in the hierarchy?

6. Discuss the advantages and disadvantages of global food chains. Discuss the advantages and disadvantages of local food systems.

REFERENCES

Baskin, Y. 1997. The work of nature—How the diversity of life sustains us. Island Press, Washington, DC.

Björkland, J., K. Limburg, and T. Rydberg. 1999. Impact of production intensity on the ability of the agricultural landscape to generate ecosystem services: An example from Sweden. Ecol. Econ. 29:269–291.

Brandt, J., B. Tress, and G. Tress (ed.) 2000. Multifunctional landscapes: Interdisciplinary approaches to landscape research and management. Proc. International Conference. 18–21 October. Centre for Landscape Research, Univ. Roskilde, Denmark.

Daily, G.C. (ed.) 1997. Nature's services: Social dependence on natural ecosystem services. Island Press, Washington, DC.

Deere, C.D. 1997. Reforming Cuban agriculture. Dev. Change 28:649–669.

Det Kongelige Landbruksdepartement (The Norwegian Ministry of Agriculture). 1999. Om norsk landbruk og matproduksjon (On Norwegian agriculture and food production). Ministry Proposition 19. Norwegian Min. Agric., Oslo, Norway.

Diamond, J. 1997. Guns, germs, and steel: The fates of human societies. W.W. Norton, New York.

Doherty, S., T. Rydberg, and L. Salomonsson. 2000. Ecosystem properties and principles of living systems for sustainable agriculture. p. 152–155. *In* T. Alföldi et al. (ed.) IFOAM 2000—The World Grows Organic, Proc. International IFOAM Conf., 13th. Frick, Switzerland. 25 Aug.–2 Sept. 2000. Vdf Hochschulverlag, Zürich.

Drake, L. 1999. The Swedish agricultural landscape—Economic characteristics, valuations and policy options. Int. J. Soc. Econ. 26(7,8,9):1042–1060.

Forman, R.T.T., and M. Godron. 1986. Landscape ecology. John Wiley & Sons, New York.

Francis, C. 2004. Soil dynamics, plant nutrition, and soil quality. p. 31–48. *In* Agroecosystems analysis. Agron. Monogr. 43. ASA, CSSA, SSSA, Madison, WI.

Francis, C., G. Lieblein, S. Gliessman, T.A. Breland, N. Creamer, R. Harwood, L. Salomonsson, J. Helenius, D. Rickerl, R. Salvador, M. Wiedenhoeft, S. Simmons, P. Allen, M. Altieri, J. Porter, C. Flora, and R. Poincelot. 2003. Agroecology: The ecology of food systems. J. Sustainable Agric. 22(3):99–119.

Francis, C., and D. Rickerl. 2004. Ecology of food systems: Visions for the future. p. 177–198. *In* Agroecosystems analysis. Agron. Monogr. 43. ASA, CSSA, SSSA, Madison, WI.

Giorgis, S. 1995. Rural landscapes in Europe: Principles for creation and management. Council for Europe, Strasbourg, France.

Gliessman, S.R. 1998. Agroecology: Ecological processes in sustainable agriculture. Ann Arbor Press, Chelsea, MI.

Kinzig, A.P., J. Antle, W. Ascher, W. Brock, S. Carpenter, F. Stuart Chapin III, R. Costanza, K.L. Cottingham, M. Dove, H. Dowlatabadi, E. Elliot, K. Ewel, et al. 2000. Nature and society: An imperative for environmental research [Online]. Rep. from a Workshop, Developing a Research Agenda for Linking Biogeophysical and Socioeconomic Systems, Tempe, AZ. 5–8 June 2000. Available at http://lsweb.la.asu.edu/akinzig/report.htm (verified 24 July 2003).

Lieblein, G., C.A. Francis, and H. Torjusen. 2001. Future interconnections among ecological farmers, processors, marketers, and consumers in Hedmark County, Norway: Creating shared vision. Hum. Ecol. Rev. 8:60–71.

Lohr, L., and L. Salomonsson. 1999. Conversion studies for organic production: Results from Sweden and lessons for the United States. Agric. Econ. 22:133–146.

Olson, R.K. 1995. Diversity in agricultural landscapes. p. 121–160. *In* R.K. Olson et al. (ed.) Exploring the role of diversity in sustainable agriculture. ASA, Madison, WI.

Olson, R.K. 1999. Introduction. p. 1–13. *In* Under the blade: The conversion of agricultural landscapes. Westview Press, Boulder, CO.

Olson, R.K., and C.A. Francis. 1995. A hierarchical framework for evaluating diversity in agroecosystems. p. 5–34. *In* R.K. Olson et al. (ed.) Exploring the role of diversity in sustainable agriculture. ASA, CSSA, SSSA, Madison, WI.

Risser, P.G. 1987. Landscape ecology: State of the art. p. 3–14. *In* M.G. Turner (ed.) Landscape heterogeneity and disturbance. Springer-Verlag, New York.

Steinmetz, S. 1993. Webster's desk dictionary. Gramercy Books, New York.

Torjusen, H., G. Lieblein, M. Wandel, and C.A. Francis. 2001. Food system orientation and quality perception among consumers and producers of organic food in Hedmark County, Norway. Food Qual. Prefer. 12:207–216.

11 Ecological Morality: A New Ethic for Agriculture

FREDERICK KIRSCHENMANN

Iowa State University
Ames, Iowa

> The astonishing thing about our deepened [scientific] understanding of reality over the last four or five decades is the degree to which it confirms and reinforces so many of the older insights of humanity. The philosophers told us we were one, part of a greater unity that transcends our local drives and needs. They told us that all living things are held together in a most intricate web of inter-dependence.... These were, if you like, intuitions drawn in the main from the study of human societies and behavior. What we now learn is that they are factual descriptions of the way in which our universe actually works.
>
> —Rene Dubos and Barbara Ward

> That we failed to learn from them [Native Americans] how to live in this land is a stupidity... they had a responsible sense of living within the creation—which is to say that they had, among much else, an ecological morality...
>
> —Wendell Berry

> Had they [European immigrants] been other than they were, they might have written a new mythology here. As it was they took inventory...
>
> —Frederick Turner

History presents us with a curious paradox concerning ethics in agriculture. On one hand, philosophers, sages, and shamans have promoted the idea that humans are part of a web of life on which our survival absolutely depends and which we ignore at our peril. On the other hand, human actions have, more often than not, contradicted that wise advice. Certainly the "failures are more numerous than the successes, as told by the ruins and wrecks..." (Lowdermilk, 1953). The unraveling of that mysterious contradiction has, of course, preoccupied philosophers and theologians for millennia.

Any effort to formulate an appropriate ethic for agriculture will need to recognize the complexity of this fundamental human paradox. While this chapter does not finally unveil the core mystery of the paradox, it does attempt to recognize and honor it. Any attempt to propose an ethical blueprint for agriculture that does not take into consideration this fundamental paradox will, at best, be an abstraction that will have little to contribute to the real world of farming.

It is important to acknowledge that the culture that informed our farming practices for most of agriculture's 12 000-yr history was wedded to a story of the universe that no longer holds true. For most of human civilization, nature was perceived as a given. It was assumed that stars and trees, animals and plants, mountains and lakes, birds and worms had always been there and would always continue to be there. It was a story of stability, of permanence, of firmness—it was the story by which we lived in the world. While we now know that such a view of the world is patently false, we still farm largely in accordance with its cultural baggage. Most farmers still believe that nature is essentially permanent and stable, and they even view changes in weather patterns as temporary abnormalities (Schneider, 1976).

Modern science evolved in that same context and became a second cultural force that shaped agriculture. Scientists saw nature as a collection of raw materials that were part of a mechanistic world available to be shaped into a permanent habitat suitable solely for humans. Following that line of thought, it was presumed that nature could be controlled by humans. For the Puritans, it was all part of "taming the wilderness" and building a "Kingdom of God." For Francis Bacon, it raised the prospect of "bending nature to our will." As science developed ever more powerful technologies, it seemed more feasible to manufacture an entire food production system completely under human control—food produced entirely by human wit.

In the latter part of the 20th century, the culture of agriculture was shaped by a third force: a money-based economy. The emergence of this exclusive money-based, laissez-faire, libertarian economic theory, and subsequent economic practices, led to a conviction in agriculture, as well as in all other human enterprises, that the market always knows best and that price alone captures appropriate value. Consequently, it led to the notion that "what is not reflected in the price system does not exist" (Kuttner, 1997). Accordingly, anything that does not capture value in the marketplace is "externalized." On the farm that translated into a culture of agriculture that ignored soil loss, water pollution, loss of biodiversity, decaying rural communities, and anything else that didn't directly affect the bottom line.

It is true that a long procession of philosophers, shamans, and sages recognized the interdependence of all species. They urged humans to stand in humble awe of the intricate interrelational character of all of life, that is, to adopt an ecological morality. Yet, from the beginning, cultural forces led to very different behavior. A few cultures (mostly aboriginal) intuitively ascertained the more complex, interrelational character of the world—a view now corroborated by evolutionary biology. But for the most part, humans acted solely to secure their own survival. Consequently, humans became a "patch disturber" species who hunted out or farmed out their neighborhoods and then moved on (Rees, 1999).

It is important to recognize here, however, that humans are not the only species to modify their environment. In an effort to secure their own survival, all organisms modify their environment, and in doing so actually *create* the environment. Were it not for organisms constructing their own environment out of the bits and pieces available to them, there would be no environment. But the way each species modifies its environment both destroys part of the environment and creates opportunities for other species in its ecological neighborhood. Each organism uses resources that are in short supply and transforms them into a form that cannot be used again by individuals of the same species, but does become a resource for other species. Each species, in other words, is destroying part of its environment, and in the process of doing so serves up resources for other species. The majority of the waste created by one species becomes food for another—hence the cyclical, the communal, the ecological character of the planet (Lewontin, 1998).

It is this discovery of the communal, dynamic nature of the planet that has created the need for a new agricultural ethic. In the old universe story, nature was a collection of stable objects that could be transformed into a permanent habitat for the sole benefit of humans, controlled solely by human-created technologies. This inspired an agricultural ethic that concentrated on just one imperative: to produce as much as possible, regardless of the cost (Thompson, 1995). This *productionist* ethic implies that any behavior is "good" so long as it increases productivity.

That ethic is now bankrupt. It utterly fails to develop an agriculture that does not degrade the very ecological and social communities on which agriculture depends. But so long as we are in the grip of a cultural vortex that

- views nature as a static, mechanistic structure which can be controlled with technology,
- assumes nature is stable and therefore largely immune to harm, and
- operates by an economy that is solely price determined,

we are not likely to abandon our productionist ethic.

At the same time our predominant *environmental* ethic, which seeks to "save" the environment, or to keep things as they are, or to maintain some kind of presumed "harmony" in nature, also is bankrupt. Nature, we now understand, is a complex and dynamic community that is in a constant state of change. Ninety-nine percent of all the species that ever existed on this planet are now extinct. And in the evolutionary journey ahead, many more species, very likely including our own, also will become extinct. While human activity, driven by a productionist ethic, is largely to blame for the current rate of extinctions, some would still occur even if there were no humans on the planet at all.

What we need, therefore, is neither an ethic that values production over everything else, nor an ethic that seeks to keep everything as it is. What we do need is an ethic that recognizes the need for agriculture to be conducted in a manner that makes a decent life for humans possible on this planet while, at the same time, retaining the ecological dynamics that sustain all life on the planet. And that will require, among other things, an agricultural ethic that respects the complex, dynamic, ecological interrelationships in which a farm exists. For agriculture to survive, it will need to honor the complex ecological mix in which it participates. In modifying the environment, the farmer needs to attend to the many and varied modifications simultaneously being carried out by the millions of other organisms on which the farm depends. Ideally, all the modifications will form a kind of synergy that increases the robustness of the entire ecological neighborhood. It will require, in other words, an ethic that upholds and promotes an ecological morality.

FROM A PRODUCTIONIST ETHIC TO ECOLOGICAL MORALITY

Moving from one values perspective to another is never simple and there are perhaps no strategies that we can implement to accomplish such transformations. Still, it is worth considering at least two classical proposals.

The Enlightenment Approach

One approach to changing our values is sometimes referred to as the enlightenment approach, favored by Thomas Jefferson. He believed that if people were free to do the right thing and had the information to do the right thing, they would do the right thing. David Orr suggested three modern strategies for using this approach to values transformation.

The first enlightenment strategy suggested by Orr invokes drama to emphasize the need to change course. We can "try to capture public imagination by dramatizing aspects of our situation," he wrote. He cited an example of this high drama in the Clock of the Long Now Foundation that proposes the creation of a 10 000-yr clock to illustrate the longer sweep of time. The clock would make us more "amenable to precautionary steps to preserve those things essential to the long now, and less susceptible to the political, technological, and economic contagions of the moment" (Orr, 2000). Orr acknowledged that such drama runs the risk of simply becoming another theme park, but he still thinks it is worth considering. An example of this strategy for transforming our agricultural values might be a dramatic Disney-type kind of display at county fairs throughout the nation that projects the continuing

- loss of soil due to erosion,
- loss of land to salinization and urban sprawl,
- development of dead zones, and
- demise of rural communities,

with an emphasis on the impact that these ongoing realities portend for our common quality of life.

A second enlightenment strategy that Orr suggested lies in "creating more accurate and telling metaphors and theories." An example of this strategy is the effort to include ecological capital in our economic accounting. When applying this strategy to the task of transforming agricultural values, one might target information that demonstrates how much additional money farmers have to spend in fertilizer costs for every tonne of soil lost on their fields, or how much more it will cost them for farm maintenance when goods and services provided by local rural communities disappear. Again, Orr reminds us that changing opinions by this method tends to be slow—mostly "funeral by funeral."

A third enlightenment strategy consists of political change. This strategy suggests that people use the machinery of the body politic to conduct the kind of public interest research and promote the robust political ideas that would protect "earth systems." Orr acknowledged that this strategy is particularly problematic in the current political climate, since our belief that the private sector is superior to government in almost every respect has dramatically weakened government.

But that cautionary tale serves as a summary of the overall problem with the enlightenment approach: One's current commitment to old values creates institutions and entrenched cultural mores that tend to prevent us from seeing the need to implement new values. Consequently, the enlightenment approach is not a very promising way to bring about a real and rapid transformation of values.

The Incentives Approach

A second classical approach to changing human behavior is by creating incentives to change. Adam Smith, the best-known proponent of this approach, believed that even if people are free to do the right thing and have the information necessary to do the right thing, they are still unlikely to do the right thing unless there is some incentive for them to do so.

According to this approach, if we want farmers to adopt an ecological morality instead of a productionist ethic, we need to develop policies and markets that reward them for doing so. In the policy arena there are examples where this approach has been implemented, such as the U.S. Conservation Reserve Program (CRP). In this program farmers are paid by the government to plant highly erodible land into grass and discontinue any agricultural practices for a period of 10 or 15 yr. Since the program's implementation, farmers have offered to put more land into the Reserve than could be accepted with available funds, demonstrating that the incentive works, albeit at a very high cost. It also is clear that this incentive achieved many of the environmental goals the program envisioned. The program vastly increased habitat for wildlife, dramatically reduced soil erosion, and improved water quality.

Unfortunately there is little evidence that the CRP has changed the values of the farmers who signed up for the program. As CRP contracts expire, most of the land has returned to production, under management practices identical to the ones that were in place prior to the contract. It would appear that incentives created by public policy are effective only in changing behavior as long as the incentives are in place.

Are market-driven incentives any more successful in achieving an ecological morality than policy-driven initiatives? First, it is important to recognize that market-driven success hinges mostly on one's intuitive capacity to anticipate new opportunities. If incentives influence human behavior only as long as the incentives are in place and market-driven incentives are dependent on the entrepreneur's ability to intuitively anticipate new opportunities, then this is not a very effective way to establish a consistent ethic. While there may be potential for additional response if market-driven incentives are sustained for a number of years, practices are likely to revert to the status quo when these incentives lapse.

The organic food industry, which emerged from an ecological ethic, offers a good example of what can happen. Sir Albert Howard, one of the architects of the organic movement, embodied the ethical beliefs that guided the movement. The health of the soil was Howard's guiding principle. For him, the central question was whether humans could regulate their affairs so that their "chief possession—the fertility of the soil—is preserved" (Howard, 1943). However, once entrepreneurs intuitively anticipated new opportunities in the organic market, a very different kind of ethic emerged. For many in the food industry, organic has now become a "corporate adventure" that, in many respects, makes the organic industry barely distinguishable from the conventional food system (Pollan, 2001).

Yet we should not overlook the power of the market and its incentives as a tool to move us toward more ecologically sound farming practices. Currently in the food and agriculture system there are many signals indicating that the market may be ripe for some changes. These shifts could spur the development of incentives that may lead farmers to adopt more ecologically sound practices.

The current market system, for example, is not working well for the environment *or* farmers. A recent survey conducted by Paul Lasley, a rural sociologist at Iowa State University, revealed that 64% of Iowa farmers believed that the overall economic prospects for farmers in the state would become worse in the next five years (Lasley, 2000). At the same time the environment continues to signal that nature's sinks can no longer tolerate the wastes expelled into the environment by industrial farming practices. A recent report on N pollution indicated that if today's enterprises, including agriculture, continue to function in their current mode, the amount of N pollution added to the environment in the next 25 yr will double (Horton et al., 2000). A steady stream of reports concerning the deterioration of Iowa's freshwater lakes and streams appeared in the *Des Moines Register* during the spring of 2001 and serves as a harbinger of that prediction. Sports enthusiasts, environmentalists, and tourists almost certainly will begin to exert pressure on farmers to change their farming practices to prevent further pollution (*Des Moines Register*, 2001).

Contemporary market analysts suggest that a growing number of food shoppers want to have a conversation about their purchases (Locke et al., 2000). This means, among other things, that consumers increasingly will want to know the full story of the food they purchase—where it came from, whether environmental stewardship was practiced in producing and processing it, and whether laborers were treated fairly, and the story had best be authentic (Schlosser, 2001).

These are all negative indicators that give intuitive signals for potential new market opportunities. If conventional agriculture is not working for farmers *either* economically *or* ecologically, then the stage is set to develop new markets for food products that are produced in a more ecologically sound manner and that can retain more value on the farm. The farmer becomes an integral part of the story surrounding food production.

But the larger problem with the incentives approach to developing an ecological morality remains. Apart from an abiding ethic that forms the context of economic activity, incentives will always shift to whatever opportunity the market finds attractive. If that happens to be damaging to local communities or local ecologies, the market doesn't care. In a libertarian market system, economics determine values, while values never seem to determine the market.

But the reality is never that simple. Economies always function inside a set of values. Markets always demand some regulations. Few want to let the market alone decide the level of food safety. No one wants to let the market decide whether or not recreational drugs should be sold freely in supermarkets. The real question, then, is what kind of values drive our market economies?

The answer to that question has become increasingly complex since the end of the Cold War. "Now, there are many different 'mixed economies' based on countries using their own markets, rules, laws, contracts, and so on, reflecting their own goals and values, traditions and cultures," according to Hazel Henderson (1997). She also stated in her book that "Today, we are rediscovering that values, far from being peripheral, actually drive all economic, technological and social systems... Today, we must clarify the immense political, social, economic, and technological transitions we are experiencing. How do these transitions relate to new and old multicultural goals and values, to ideas of wealth and progress, satisfaction, freedom, and development, and to the deeper spiritual and religious concerns they embody?" In other words, not only are we developing a money-based global economy, we also are developing a global civic society that will be engaged in sorting out these very complex values issues. Today we can see the tension between those two forces playing out on the global stage. Certainly the global civic society will not go away. The question is, will an ecological morality be part of the new global culture that evolves out of this process?

In the end, the extent to which we move toward an ecological morality will depend on a process that cannot be reduced to either incentives or enlightenment strategies. Rather, the outcome will be determined by a complex host of actors on the global stage, and both incentives and enlightenment will likely be part of that process.

THE PROBLEM OF ETHICAL DISSONANCE

This brings us back to our opening observation. We humans often act contrary to our own rational self interest and, even more so, contrary to our own value commitments (Kirschenmann, 2000). Consequently, if we want to create an ecological morality that actually results in farming practices that respect the dynamic interrelationships of all the organisms that make up the ecologies in which we farm, then we have to come to terms with this paradox.

Of course we are all aware of the problem of ethical dissonance. Our own communities, as well as the history of human literature, are replete with examples. But modern western culture has tended to deal with the problem of ethical dissonance by externalizing it into a kind of morality play. Classic western movies, while they were seldom allegorical or given to providing explicit ethical instructions, were nevertheless morality plays of sorts. They epitomize the way the western world deals with ethical dissonance; morality plays divide the world up between the virtuous and the wicked, the good guys and the bad guys, white hats and black hats. The good guys always win and the bad guys are always subdued in the end, often through force and violence.

This is a neat way of dealing with the messy complexity of ethical dissonance without having to honestly explore all aspects of the problem. It oversimplifies the problem, externalizes it, and exempts the viewers of the drama from having to deal with the dissonance within their own experience. Most importantly, it leaves the real world problem unsolved because each party always identifies with the side of virtue, without a real examination of his or her values and behavior.

Modern agricultural ethics provide the stage for a similar morality play where environmentalists and farmers are pitted against one another. Environmentalists feel that they wear the white hats while farmers wear the black hats. Farmers pollute and environmentalists are attempting to stop pollution. The farmers, in the meantime, see environmentalists as wearing the black hats. Farmers see themselves in white hats, trying to feed the world, while environmentalists are preventing them from doing so by imposing unworkable regulations, regulations that are, in their view, based largely on a total ignorance of farming systems (Kirschenmann, 1988). In the last analysis, this way of dealing with ethical dissonance leads to a kind of ethical constipation. Once we have conceptualized the world in this way, divided between good guys and bad guys, no real change can take place. The good guys absolve themselves of any need for change and the bad guys are never expected to change—the only way to deal with them is to subdue them. It is an approach that generally leads to denial, self-righteousness, hypocrisy, and inertia.

In his delightful little book, *The Genesis of Ethics*, Rabbi Burton Visotzky suggested an alternative approach to moral development that provides a potentially more effective way of dealing with ethical dissonance. Visotzky reminds us that the book of Genesis in the Old Testament hardly presents itself as a moral book to the modern sensibilities. As he says, it reads more like a soap opera about a dysfunctional family than a moral text. "It is a story about rape, incest, murder, deception, brute force, sex, and blood lust. The plotlines and characterizations of Genesis are so crude as to call into serious question how this book became and remained a sacred, canonical text for two thousand years and more" (Vitsotzky, 1996). In other words, it is the antithesis of a morality play.

What is going on here? How can this be a book for moral development? Visotzky argues, correctly I think, that we can identify with it precisely because Genesis spares no punches in telling its story about the human family. "It is the unattractive component of Genesis that causes us to have such a strong identification with it in the first place. When we read of the dysfunctional family with strong lust and murderous intentions, we recognize that it is our family.... " In other words, instead of providing us with a mechanism for avoiding our own ethical dissonance, as the morality play does, the Book of Genesis leads us to own our personal ethical discord.

The point is that we cannot effectively deal with ethical dissonance until we adopt it as our own condition. Only when we recognize that our own ethics are dysfunctional are we in a position to experience any kind of meaningful ethical transformation. Until we acknowledge that we are all part of the problem, there is little chance for a meaningful cultural transformation that will lead to actual changed behavior. Until farmers and environmentalists recognize that they are both part of the problem, they won't be able to work together to find real solutions that both can embrace. It is unrealistic to expect that farmers, given their present circumstances, will change their farming practices from producing as much as possible regardless of the ecological or economic cost. They are caught in a food and agriculture system and public policy structure that provide them with few options to improve their economic position—and more production is presented as their primary option. If we see a contradiction in that behavior, it is no less a contradiction than the environmentalist who burns fuel to fly across the country to join a protest against farming practices that cause pollution. We are both part of the same system that leads to one degree or another of the same disparities. This is not to deny that some behavior is more harmful to local ecologies than others, but understanding our complicity is important.

Our approach to ethics prevents us from seeing our own dissonance, and therefore prevents us from appreciating the contradictory situations in which we all find ourselves. None of us "walk our talk," and it might be destructive if we did. Our collective failure to recognize that none of us "walk our talk" causes us to marginalize everyone in whom we see ethical contradictions. When the media noted that protesters involved in the "Battle of Seattle" were wearing Nike shoes, the news people dismissed the protestors as being insincere and therefore not worthy of our attention. When we discover that an environmentalist living in the mountains owns an SUV, we don't believe that her message can have any value worth our notice. At the same time, we ignore the contradictions in our own lives.

The Genesis story therefore has great insight to offer to our own situation. Until we recognize the foibles and foolishness in our own lives, we are not likely to participate meaningfully in the development of a new ecological morality. Unless farmers and nonfarmers alike recognize their common need to rethink core values, we are not likely to see the evolution of a new ethic for agriculture that reflects our new understanding of how nature works. We need to participate fully in a robust living community in which we recognize that our participation in nature is one small part of a living web that includes millions of other organisms as our neighbors and fellow creators of the environment in which we all have a communal stake. And we need to honor the fact that our farms are part and parcel of that ecological neighborhood. From that perspective we may still have an opportunity, for a short while, to write a "new mythology here" instead of taking inventory simply to ascertain how much of nature we can use in a futile attempt to shore up an isolated habitat for humans on the planet (Pollan, 1996).

STUDY QUESTIONS

1. Is it reasonable to assume that we can "save the environment"? Why or why not?

2. Is technological innovation likely to solve the "environmental crisis"? Why or why not?

3. Is better education likely to solve the "environmental crisis"? Why or why not?

4. Are appropriate incentives likely to solve the "environmental crisis"? Why or why not? Describe some examples of such incentives.

5. Do you think farmers and environmentalists will ever be able to work together for a common future? How might that happen?

6. How might the sciences of ecology and evolutionary biology inform our ethical choices with respect to agriculture and the environment?

REFERENCES

Des Moines Register. 2001. Various articles and op ed pieces. 6–22 May.

Henderson, H. 1997. Building a win–win world: Life beyond global economic welfare. Berrett-Koehler, San Francisco.

Horton, T., H. Dewar, and F. Langfitt. 2000. The Baltimore Sun. 24–28 September.

Howard, Sir A. 1943. An agricultural testament. Oxford Univ. Press, New York.

Kirschenmann, F. 1988. Resolving conflicts in American land-use values: How organic farming can help. Am. J. Altern. Agric. 3:1.

Kirschenmann, F. 2000. Challenges facing philosophy as we enter the 21st century: Reshaping the way the human species feeds itself. Willard O. Eddy Lecture, Colorado State University. Available from the author.

Kuttner, R. 1997. Everything for sale: The virtues and limits of markets. Alfred A. Knopf, New York.

Lasley, P. 2000. Iowa farm and rural life poll: Summary report. PM 1801. July. Iowa State Univ., Ames.

Lewontin, R. 1998. The triple helix. Harvard Univ. Press, Cambridge, MA.

Locke, C., et al. 2000. The cluetrain manifesto: The end of business as usual. Perseus Publ., Cambridge, MA.

Lowdermilk, W.C. 1953. Conquest of the land through seven thousand years. Agric. Info. Bull. 99. USDA, Soil Conservation Service, Washington, DC.

Orr, D.W. 2000. Ideascleroisis: Part one. Conserv. Biol. 14:4.

Pollan, M. 1996. Second nature: A gardener's education. Dell Publ., New York.

Pollan, M. 2001. Naturally: how organic became a marketing niche and a multibillion-dollar industry. The New York Times Magazine. 13 May.

Rees, W. 1999. Scale, complexity and the conundrum of sustainability. *In* M. Kenny and J. Meadowcroft (ed.) Planning sustainability. Routledge, New York.

Schlosser, E. 2001. Fast food nation. Houghton Mifflin, Co., Boston, MA.

Schneider, S.H. 1976. The genesis strategy: Climate and global survival. Plenum Press, New York.

Thompson, P. 1995. The spirit of the soil. Routledge, New York.

Vitsotzky, B.L. 1996. The genesis of ethics. Crown Publishers, New York.

12 Ecology of Food Systems: Visions for the Future

CHARLES FRANCIS

University of Nebraska
Lincoln, Nebraska

DIANE RICKERL

South Dakota State University
Brookings, South Dakota

In the introduction we invoked artistic renditions to convey a sense of holism and systems at different levels of scale, dimension, and complexity (Chapter 1, Rickerl and Francis, 2004). Agroecology as an area of research and education is increasingly viewed as encompassing multiple dimensions that extend across different levels of spatial scale. Some educators define agroecology broadly as the ecology of food systems (Francis et al., 2003). It is too soon to determine if this expanded definition will become common currency in research and education, yet it is beyond doubt that agroecology encompasses the wide range of perspectives presented in the previous chapters.

As illustrated in the chapter on multifunctional landscapes (Fig. 10–1, Chapter 10, Francis et al., 2004), agricultural systems may include ecological, economic, social, ethical, and legal aspects that influence choice of enterprises, farming practices, farm design, and products sold. Beyond the farm gate, many decisions drastically impact the economic success of an individual farm. Many of the factors that affect food security operate in the marketing, international finance, and even political arenas. The multiple dimensions of these factors, and the decisions and potentials for success in farming, are operative at different spatial levels, a concept that in this book has emerged as one of the bases for explaining performance of agricultural systems. To the extent that we can identify different levels in the hierarchy where the action takes place, and which are the most important factors at each level, this understanding can contribute to a research agenda for the design of new and more efficient and profitable systems. We too often work at the plot or field level in agronomy, for example, and expect to extrapolate the results to the landscape and region without contextualizing the information. To the extent that we can bring an artistic eye to design of the farm and landscape, we are capable of envisioning new combinations of enterprises that interact in unique ways, creating emergent properties that may enhance the performance and efficiency of each system.

At the 2001 ASA symposium on defining the agenda for agroecology, speakers emphasized the importance of exploring biogeochemical interactions in the design of cropping systems and how to present this concept to farmers (Harwood et al., 2001), as well as the importance of communities and human capital in relating farming systems to the broader society (Flora, 2001b). The practical applications of agroecology in an Extension framework are well presented in the series of publications from Michigan State University (e.g., Cavigelli et al., 1998). The complex interactions between farming systems in the rural landscape and their human communities were explored by a number of authors in a recent book (Flora, 2001a). It is obvious from the presentations in the ASA symposium in Charlotte, NC and in recent publications that future study of a complex field such as agroecology must of necessity be an interdisciplinary activity that requires us to move beyond a department and disciplinary mind set. This perspective informs our focus on broad spatial and temporal scales of interest, as well as the need to involve a wide range of actors on the stage of research and education.

Just as important as moving farming from a monoculture cropping system to a diverse farm and landscape, it is essential that we broaden our perspectives on research and education from the "monoculture of the mind" that currently dominates and maintains many programs in the mainstream. The process of envisioning change and exploring why it is difficult was studied by Thomas Kuhn (1962) in *The Structure of Scientific Revolution*. He argued that people invested in one paradigm have great difficulty seeing another, as did the blind men seeing the elephant in our analogy in Chapter 1 (Rickerl and Francis, 2004). In fact we may be blind to evidence that suggests systems should move in a new direction when that evidence is contradictory to the prevailing wisdom in a particular field. Kuhn's concepts have been popularized and put into practical application by Joel Barker (1985) in his several books and video presentations. Another visioning milestone in industry was achieved by a consultant, Marjorie Parker, working with the large fertilizer corporation Norsk Hidro in Norway (Parker, 1990), where a broad consensus of ideas and opinions was generated by tapping into the creative energy of all employees at all levels in the business. In each of these books, the authors urge us to go beyond our established norms and the expected ways in which systems perform. Their advice leads us to heed the oft-quoted statement of Nobel laureate bacteriologist René Dubos, "Trend is not destiny."

In academia, there is an opportunity to learn from approaches to teaching and research being taken by Tribal Colleges and Universities that in1994 were made part of the Land Grant System. The Land Grant philosophy initially embraced education for the rural poor and focused on agricultural research to benefit farmers, ranchers, and rural communities in each state. The approach taken by many 1994 Land Grants has been a return to this focus on serving the community. In Tribal colleges and universities surveyed in South Dakota, serving the community was unanimously listed as the most valued goal of research and education. Respondents also pointed out that "Our [1994 Institutions] research model might not look like yours in the original land grant universities."

To summarize the ideas in this collection of chapters we examine alternative approaches for the future, viewed across a hierarchy of spatial scale. The multiple dimensions and multifunctionality of agroecosystems and rural landscapes must be recognized to provide context for research and education. An important dimension of these multiple functions includes the ecosystem services that rural landscapes provide. To put programs into action and reach meaningful results in this complex research and education process, we advocate more emphasis on systems studies. Especially important are those programs that allow our vision to take in the entire farm and landscape and beyond, and to explore measures that expand our normal "yield per acre and net return per year" mentality. It is also essential that we continue to explore the mechanisms of agro- and natural ecosystems to better understand how they work and how we can design productive and sustainable alternatives for the future. We urge researchers and educators to view these components and mechanisms as operating in concert within a larger system and to know that to understand their relevance it is essential to view all results in the context of both biological and social realities. Types of analyses appropriate to the study of systems are explored beyond the typical enterprise aggregation and whole-farm budgets that generally characterize this type of accounting. The information from prior chapters is examined across the spatial hierarchy, and future visioning is applied to design of a research and education agenda that will be appropriate for complex and rapidly evolving agriculture and food systems. Different and more robust methods of analysis are suggested as integral to this process. We conclude with a series of suggestions for future research in our areas of crops, soils, and agricultural systems, and how these can be integrated with social sciences in exploring the complexity of how food is produced and moves to the consumer. We are convinced that agroecology provides a number of useful methods and guidelines for design of future systems.

MULTIDIMENSIONALITY OF AGRICULTURE AND FOOD SYSTEMS

Farmers, researchers, and educators appreciate the increasing complexity of agriculture and food systems, while consumers in today's society become ever more distanced from their sources of food. From the consumer point of view, a system that provides choices of 50 types of breakfast cereal, accompanied by citrus or bananas every day of the year (in the U.S. supermarket), leaves little to worry about except for how to pay the bill (Schlosser, 2001), and when food constitutes only 11% of the average U.S. family budget, even economics of food are of negligible importance to the majority. Although the average consumer knows that the raw materials for food in the supermarket somehow come from agriculture and rural areas, they probably do not know where each food item comes from nor the conditions under which it was produced. In addition, many producers know little about processing beyond the farm gate. New ideas for value added agriculture and locally grown foods need to integrate producers and consumers, not unlike the model offered in a natural ecosystem. Even less apparent are the impacts on rural landscapes and communities, long-term consequences for the environment, profitability for farmers, and social implications for farm labor here and elsewhere. It is within this context that farmers must make a living with high production costs and low commodity prices, researchers must attempt to improve agriculture while faced with lagging budgets, and educators must try to help students discover how the system works. Only by recognizing the multiple dimensions of agriculture and food systems can we hope to make any sense of the situation and take a rational look at the future.

Dimensions of interest in agriculture and especially the social components are often narrowly defined. Discussion of resources in the rural environment at any level in the spatial hierarchy has focused on the soil, rainfall, crop varieties and hybrids, and purchased inputs that lead to production and profit for the farmer. Economic resources that allow farmers access to the land, labor, and inputs are among the determinants of how successful a given farm will be and how willing people will be to consider innovations in crops, practices, or systems. Most research in agricultural science is focused on system components. However, when Lakota Elders at a gathering on the Rosebud Reservation in South Dakota were asked by university scientists to prioritize agricultural research areas, they did not understand how natural resources, food production, and community health could be separated in order to be prioritized. Collaboration with cultures that have a more holistic view is teaching us about multidimensional thinking in agriculture.

Human resources, or rural social capital, include the farmer or manager, a key person who represents the focal point where decisions are made and future directions are set for individual farms. Decisions frequently depend on the availability of operating capital, and bankers tend to be conservative and comfortable with existing crops and systems. Education, experience, current economic situation, and perceptions of alternatives are all part of the cultural and social environment in which these decisions are made. And the rural community is the place where components of social capital come together to create social control, a topic explored in detail by Cornelia Flora in Chapter 7, Community Dynamics and Social Capital (Flora, 2004). Along with the multiple functions of rural landscapes, those dimensions are often simplified or even erased by industrial agriculture. The social implications of current agriculture and land ownership must be seriously approached if we are to find sustainable systems that will serve people. These are among the most important dimensions of agriculture and food systems that must be considered in design of the future.

For research to have a real impact on productivity, resource use, and sustainability of agriculture, we must conduct our experiments on systems that will make the greatest difference and do this at the appropriate level of scale. Further, we must frame that research including interpretation of results within the context of the whole agroecosystem. There is certainly a need for testing new crop hybrids and alternative sources of nutrients for plants, but these tests should incorporate an evaluation of the impacts of the various treatments on production, economic returns, long-term environmental impact, and social implications for the family and community. Just calculating yields or the economic bottom line for a single season is not sufficient. Nor can we assume that higher production on the farm, positive net income for a specific commodity, or a positive economic bottom line for the farm at the end of the year will necessarily lead to a greater positive human impact of this success in food production. Too many other factors impact the overall effects of changes in technology, and these are among the multiple functions of agriculture that need to be a part of research and education in agroecology (Francis et al., 2003).

IMPORTANCE OF ECOSYSTEM SERVICES

Beyond the apparent biological, natural, and human resources in the rural environment and agroecosystems that easily fit into our taxonomy, department structure, and current descriptions, there are less obvious interactions and emergent properties that provide ecosystem services and greatly benefit society. The importance of interactions among factors was discussed in Chapter 10 by Francis and colleagues from the Nordic Region (Francis et al., 2004), with major and minor interactions between components of natural systems and those of social systems shown at a single spatial level (Fig. 10–2) and between components across levels of scale (Fig. 10–4). Some consequences of what emerges from these complex combinations and interactions of factors are a number of vital ecosystem services.

Daily et al. (1997) characterized these services as (i) "essential to civilization," (ii) operating "on such a grand scale and in such intricate and little-explored ways that most could not be replaced by technology," (iii) already highly impacted and impaired by human activities, and (iv) susceptible to being "dramatically altered in all of Earth's remaining natural ecosystems within a few decades." Their list of services relevant to agroecosystems, farms and families, and rural communities includes the following:

- Purification of air and water
- Mitigation of droughts and floods
- Generation and preservation of soils and renewal of their fertility
- Detoxification and decomposition of wastes
- Pollination of crops and natural vegetation
- Dispersal of seeds
- Cycling and movement of nutrients
- Control of the vast majority of potential agricultural pests
- Maintenance of biodiversity
- Protection from the sun's harmful ultraviolet rays
- Partial stabilization of climate
- Moderation of weather extremes and their impacts
- Provision of aesthetic beauty and intellectual stimulation that lift the human spirit

These same services are also important for urban dwellers. Our society, urban and rural alike, often undervalues such services, or in fact we do not recognize them at all. The rural landscape and the managed agricultural systems we implement, herein called agroecosystems, are highly impacted by our decisions on which types of farming, the choice of practices, and the degree of conservation that we are able to implement. Much of the quality of life of people in both rural and urban areas depends on these decisions, and they reflect on the multiple activities and outcomes from the agricultural and food systems that we employ. Daily et al. (1997) pointed out that it is not surprising that we take these services for granted, since most of them were in place and in balance far before our species appeared on the scene. In general, the erosion of such services has been a gradual rather than a spectacular process, outside of major flood or fire events, or phenomena such as the Caribbean Dead Zone. Carefully directed research is needed to search out appropriate methods to understand both disruption and sustainability in the system as we evaluate systems and technologies, even as we recognize that our ability to understand and evaluate complex impacts is limited by our methods and abilities to focus on larger levels of scale.

SYSTEMS RESEARCH AND METHODOLOGY

One of the most powerful influences over the past several decades to raise awareness of the connectedness of components and importance of looking at whole systems was Rachel Carson's (1962) *Silent Spring*. Beyond the well-publicized examples of DDT weakening shells of bird's eggs and its disastrous effects on reproduction, Carson explored the complexities of biological systems and the potential disruption that a range of chemicals and their mixtures could have on function of individual species and ecosystems. Four decades later, we are still finding new scientific evidence that supports her thesis. This example was brought home in our International Agriculture course when students traveled to Canada and New Zealand. Researchers in Montreal were investigating the introduction of DDT into the food chain of several indigenous groups in northern Canada. In New Zealand, farmers who wanted to diversify their operation to include dairy were stopped by regulations on DDT levels in pasture soils. Although the former group had never applied DDT, and the latter group had overapplied this insecticide, both faced food web implications and a decline in diversity. Rachel Carson's work provided a strong incentive to examine whole systems, an incentive that is relevant today.

Gordon Conway (1985) was an early advocate of bringing a systems perspective to research in agriculture, through a case study approach that included systems evaluation using the methods from several disciplines. Smolik et al. (1994) combined case study and research plot approaches in systems studies including economic, environmental, and energy interpretations. The experience in systems research and education at Hawkesbury in New South Wales has been considered by many as a groundbreaking approach to integration of critical components and their interpretation within a farm context (Bawden, 1991). Widely quoted references on agricultural systems are those by Spedding (1979, 1988), who looked at carefully defining systems and the types of change or improvement that were desired and how to measure them. In contrast to the biological and physical science approach, Collinson (2000) summarized the work of two decades in farming systems research and extension that was built on in-depth conversations with farmers and interventions that closely involved both farmers and development groups in the process of system improvement. These field activities in the developing world were practical applications of Checkland's (1981) blending of "hard systems" and "soft systems" approaches, roughly corresponding to the methods used in biological and physical sciences with those used in social sciences.

Recent approaches in research and education have evaluated agricultural and larger human systems in terms of land, fossil fuels, and other resource use. Alternative valuation methods were summarized by Doherty and Rydberg (2002), and these have been applied to agroecosystems and food systems analyses by Francis et al. (2003). The environmental footprint (Wackernagel and Rees, 1996) has been proposed as a measure of the balance of carbon emission and absorption by all of modern society's activities, an index that suggests the USA (lower 48 contiguous states) currently gives off about six times as much CO_2 as that potentially absorbed by all the green plant activity in the same area. Emergy analysis (Odum, 1996) is proposed as a measure of both the quantity and quality of energy used by individual activities or industries, another way to evaluate the efficiency and impact of energy-using systems. A valid measure of the entire energy and materials efficiency of alternative products can be calculated by the life cycle analysis of individual items (Audsley et al., 1997), a method that considers costs of the infrastructure for producing and for using an item as well as the costs of its disposal once the useful life is finished. Finally, the concept of a comprehensive checklist that could be used to evaluate sustainable landscape management was developed by a European consortium of scientists (van Mansvelt and van der Lubbe, 1999). They include criteria related to environment, ecology, economy, sociology, psychology, and physiognomy (the objective integration of landscape elements). This section provides an overview of some of the tools available for research in agroecosystems, in addition to those included in the earlier chapters. Next we present an expanded list of potential priorities for research and education for the future, taken from previous chapters and from the literature.

FUTURE RESEARCH AND EDUCATION PRIORITIES

Potential future research priorities are presented across a spectrum of spatial scale, and the issue areas generally correspond to the topics covered by the chapters of this book. The discussion here is not carefully referenced, since many of the key information resources are already cited in previous chapters. As suggested by Olson (1995), we can explore mechanisms and try to understand better the functions in detail by doing precise research at a lower or more fine-tuned level in the hierarchy of scale. At the same time, research conducted at any level in the spatial hierarchy should be carefully interpreted within the context of higher levels to provide meaning and relevance to the work. We should also be careful to seek applications for research at the same level at which studies have been conducted. In education it is important to establish an appreciation of how and why we study different activities or processes at different spatial levels, and how information from one level informs our understanding of what happens at lower or higher levels. In general, specific disciplines examine agroecological processes and decisions at levels where their practitioners have experience and comfort with available research methods. For example, the primary work of agronomists is at the plot and field level; economists at the enterprise, farm, or national levels; and sociologists at the farm and community levels. One great value of interdisciplinary studies is a growing appreciation of methods that are used in areas outside our own disciplines, and the combinations of methods that add richness to the research process. The discussion of priorities is summarized by topics or issue areas in Table 12–1.

Soils and Nutrients

Nutrient dynamics are discussed in several chapters and provide the major discussion focus by Francis (Chapter 3, Francis, 2004) on soil processes related to

Table 12–1. Future research and education priorities in key issue areas organized within a special hierarchy of scale.

Issue area	Field	Farm	Landscape	Region	Continent	World
Soils and nutrients	Site specific management	Enterprise complementation	Nutrient movement across boundaries	Most efficient use of key soils	Soil and nutrient losses	Dispersion of some critical nutrients
	Alternative nutrient sources	Nutrient cycling among fields	Complementarity across farms	Widespread soil conservation	Energy invested in crop nutrients	Concentration of other nutrients
Water, sunlight, CO_2	Efficient crop resource use	Enterprise location on farm	Trapping, storing water for crops	Competing uses for water in region	Population balance with resources	Global warming due to excessive CO_2
	Cropping system design	Resource cycling and efficient use	Most efficient use of terrain for crops	Regionalization of food production	Changes in crop adaptation	Population/resource distribution
Pest management	Resistant crop varieties	Rotations of crop enterprises	Multifarm strategies for control	Regional pest dynamics	Multiregional alliances for control	Global trade group decisions on GMOs
	Integrated cultural practices and system design	Animal systems and crops rotated	Integrated pest management	Broad-based decisions, controls	Trade group talks on rules for residues	Quarantines on species movement
Environmental impacts	Soil loss requires nutrient replacement	Species diversity loss on farms	Loss of ecosystem services for society	Flooding, reduced water quality	Silting rivers, lakes, loss of habitat	Ecological dead zones in ocean
	Chemical pesticides restrict rotations	Loss of wetlands, other key elements	Loss of key habitats and ecosystems	Major costs for remediation	Disappearance of best farmlands	Global warming due to excessive CO_2
Economic impacts	Profitable crops in excess production	Reduced risk through diversity	Local markets and financial resources	Bioregional food systems for future	Federal supports distort markets	Overproduction of commodities
	Short-term decisions not sustainable	Enterprise complementarity	Viable local economies essential	Rural regions have highest poverty	Nonequitable trade agreements	Distribution of food not appropriate
Social capital	Education needed on new practices	Rural infrastructure essential for family	Rural communities essential for family	Outmigration of next generation	Mass human migration to cities	Ineffective global trade agreements
	Incentives for conservation needed	Younger farmers need incentives	Loss of cultural landscape	Rural regions have highest poverty	Migration across international borders	Inequities in food, incomes, health
Ethical issues	Plant vs. animal production/products	Farm tenure needed across generations	Lack of farm–urban connections	Loss of farming families, power	Ethical statements do not equal decisions	Food, wealth not equitably distributed
	Humane treatment of livestock	Younger farmers need incentives	Loss of rural cultural identity	Rural regions have highest poverty	Nonequitable trade agreements	Global food system does not benefit all

the production of field crops. The practical series of extension guides from Michigan State University provides an accessible window for farmers and students to grasp the importance of nutrient cycling in a field agroecosystem. Most research has been targeted at the field or crop level, and this is where most decisions are made that lead to efficient nutrient use or to excessive applications that reduce profits and could lead to environmental problems. These are all researchable areas:

At the field level, site-specific management using GPS technologies is one of the emerging areas that may catalyze efficient nutrient placement and use in the future. A logical extension of careful soil sampling and good common sense, this is really a practice that has been commonplace by family farmers for decades. People who manage a small number of fields and acres often know their land intimately and adjust their nutrient applications intuitively or with the help of soil tests or careful observations. The new technology makes some of this knowledge available to large farming operations. There is much to be done with calibration of equipment and fine-tuning the interpretation of soil tests, yield differences, and nutrient applications, and site-specific management. More exciting for those in agroecology is the growing interest in alternative sources of nutrients. Appropriate applications of manure from feed lots and confined animal operations, grazing livestock that disperse manure, and incorporation of legumes and other cover crops all hold promise for providing inexpensive crop nutrients and helping improve the environment.

At the farm level, there is research needed on the efficient complementation of enterprises, such as the design of appropriate multiple cropping and cover crop systems that occupy the land during more of the year and crop–livestock systems where both components benefit from presence of the other. Crop rotations, skillful design of enterprise diversity, and strip cropping alternatives that promote nutrient cycling among crops and animal species can reduce production costs and introduce more stability into the nutrient environment.

At the landscape level, the same type of complementarity among enterprises can be achieved. An example would be cooperation between a grain and forage producer and a neighbor who specializes in feeding out cattle. The more closely linked these activities, the greater potential for an emergent system that efficiently uses nutrients. Designing a nutrient budget across a landscape would provide information about what nutrients are needed, where they can be supplied, and how a multifarm plan could make optimum use of available internal resources. For example, in Chile the grazing of pastures and crop residues across a hillside may be followed by putting the cattle at the top of the slope to spend the night; their nutrients in manure and urine gradually move down the hillside with rainfall runoff that is captured by contour-planted crops downslope.

At the regional level, soil maps and land use categories can be used more effectively to place enterprises on the landscape, using the encouragement of financial incentives or the linking of prices to land that is farmed and grazed using the most conservation-oriented practices. Widespread soil conservation through reducing tillage, contour farming, and planting of more perennial crops can help hold soil in place and reduce or eliminate most erosion. This contributes to local soil fertility as well as reducing societal costs due to flooding and destruction of property. Permaculture options need to be researched.

At the continent level, there are also major nutrient losses that must be replaced to maintain crop productivity. Conservation efforts can reduce the need for import of nutrients or expenditure of scarce capital resources to buy energy and produce fertilizers. Energy thus saved could be used for other sectors of society or better yet conserved for future use in essential industries that have no other realistic options.

At the global level, there is concern about the dispersion of some nutrients and concentration of others. For example, P is essential for high-energy bonds that are formed through photosynthesis, and naturally occurring P in high concentration is present in only a few locations (e.g., Florida, Morocco). In modern agriculture we disperse this concentrated P on production fields, and it leaves those sites as grain that makes its way through the food chain to landfills or erodes when attached to soil particles and eventually arrives in the ocean to form nodules on the sea floor. This process of concentration to dispersion to inaccessible concentration is one that has been studied at the landscape scale. Rickerl et al. (2001) found that the dispersion of P on cropped fields and subsequent inaccessible concentration in wetlands could be reduced with the introduction of wetland buffer strips. The strips reduced nutrient loading to the wetland by almost 50%, provided hay and forage for farm animals and habitat for wildlife, and increased vegetative diversity in cropped fields. In addition, agricultural policies provided the economic incentive to establish the buffer strips. The process can be studied at the global scale for the discovery of more efficient P use alternatives.

Water, Sunlight, and Carbon Dioxide

These essential elements for crop and livestock production are discussed in several chapters. The relative advantage of agriculture is the capture of these resources across an extensive area and their conversion into plant and animal biomass that is useful for food and other products. In fact, our industry adds less value than most other sectors to concentrated amounts of these resources, and our only comparative advantage lies in the extensive nature of their occurrence and the ability of crops or livestock to capture and concentrate them. We can identify a number of research priorities at different spatial scales.

At the field level, most research has been done on plant physiology and seeking more efficient use of these often limiting resources for plant growth. Though less dominant, there has been agronomic research on how to conserve and use water by change of tillage practice or choice of crops and genotypes. Important for the future are cropping sequences and systems that capture and store water for later use, biodiverse mixtures in the field, and complementarity of intercropped species to use these resources through much of the year and convert them efficiently to useful products.

At the farm level, there is potential to store water in some areas for use in others through contour planting and directing of runoff, judicious construction of farm ponds, and placement of corrals and livestock facilities where manure can be efficiently returned to crop land. Maximum cycling of water within the farm can be accomplished by design of a diverse farmscape and choice of minimal tillage activities. Maximum sunlight interception and conversion can be accomplished by keeping a green cover over the entire farm for as many months of the year as possible, a move that will enhance capture of CO_2 and carbon sequestration as well.

At the landscape level, the efficiency of resource use will be an aggregate of the farm practices and designs included in each landscape. The placement of crops across the landscape for most efficient use of available water would require a rigorous and participatory community planning activity and some set of rewards and collective benefit from this level of organization. A method of marketing that pooled diverse products from the watershed, landscape, or community and that depended on including a wide range of adapted crop and animal species could provide the needed incentives.

At the regional level, there is competition for water currently used for agriculture from communities, industry, recreation, and environmental interests. Many of these economic activities can afford to pay more for water than we can in agriculture, and they traditionally hold first right to water use. In South Dakota, ranchers from the drier regions west of the Missouri River, who graze bison on extensive acreages, are collaborating with farmers "East River" who produce high quality hay. The idea is that if the bison cannot move to places with good grazing, the hay can be moved to them, and the goal is to buffer both enterprises against the environmental and economic damage of drought. Again, the advantage of crop and livestock production is the capture of water and sunlight across extensive areas and their concentration into useful biomass. Their efficient use compared with other options needs to be studied in that context. There are conflicts between neighboring states on the use of scarce water, for example, the ongoing discussions among Colorado, Wyoming, and Nebraska about water rights for the North Platte River.

At the national or continental level, there is concern about water use agreements and potential sale of water from one country to another vs. the rights of local populations to use and benefit from that resource. Research on the options and opportunities can lend rational information and thought to a discussion that is often a highly emotional and political battle.

At the global level, there is concern about the privatization of water systems and locking up this natural resource in multinational corporations. There is already a strong backlash in India, Brazil, and elsewhere against the sale of public water companies to international consortia, which was motivated by the goal of upgrading services but resulted in higher rates and discrimination against the poor. Research into benefits of local ownership and management is badly needed. Global climate change, especially increases in atmospheric CO_2 and the subsequent changes in crop adaptation, is under serious study. Most thoughtful scientists endorse the research that shows unidirectional change to higher atmospheric CO_2, at least in the short term, and the need for international treaties to limit emissions of CO_2 and the purchasing of carbon credits.

Pest Management

There are estimates that 20 to 40% of all crops produced worldwide are lost to pests, either in the field or during harvest and subsequent storage and handling. The rate is higher in tropical countries where both climate and infrastructure complicate production and storage of crops. For example, in the mountains of rural Bolivia, it is estimated that 40% of the indigenous population suffer from protein deficiency. Pulse crops that can grow in the area and would improve nutrition are often not produced because insects limit seed storage, and money for purchasing seed is not available to the average person. Nicholls and Altieri (Chapter 4, Nicholls and Altieri, 2004) address the dynamics of pest management at the field and farm levels. There are additional research needs and potentials to solve these challenges at wider geographic levels, and following are some priorities.

At the field level, integrated pest management (IPM) has been hailed as one of the successes of systems research and application of multiple technologies. Genetic resistance, crop rotations, clean culture, and low uses of pesticides when absolutely needed have been used effectively to minimize both costs and pest damage on crops. Further refinement of system design can enhance even more the impacts of IPM, and its use without pesticides is a cornerstone of organic and biodynamic crop production.

At the farm level, rotations of crop enterprises among fields, introduction of diversity, and crop–animal systems all have promise to help control insect, weed, and pathogen pests. Careful crop scouting to determine the degree and intensity of any pest problem can provide information needed to design improved management strategies.

At the landscape level, there can be multiform strategies to minimize pests. Synchronized planting to avoid potential host crops at different ages could benefit all participants. Increased diversity through the placement of trap crops and nonsusceptible species where pests are most likely to occur and planning of these strategies across farm boundaries in a watershed could help reduce and manage pests at the landscape level. More research is needed on pest populations and dynamics.

At the regional level, there may be need for quarantine of certain crops and products to prevent the spread of pest populations. The study of dynamics across a region will help farmers understand the sources of pests, especially insects and pathogens, and what strategies would be most appropriate for control. Regulation on what crops can be planted and when could help reduce wide-scale pest attacks and crop losses.

At the continent level, some regionalization of production can prevent serious outbreaks of pests. For example, cotton (*Gossypium hirsutum* L.) is planted in the Cauca Valley of southwest Colombia in one rainy season and on the north coast in the other rainy season; all fields must be planted within a short time frame to prevent the buildup of cotton boll weevil (*Anthonomus grandis* spp.) and make management more efficient and less costly. Agreements among participants in regional trade organizations can help establish uniform requirements for pesticide residues, and communication of research results can help everyone benefit from research in the area.

At the global level, international trade decisions on use of GMO crops that are used for insect control, quarantines on movement of seed and plant products, and agreements on pesticide residues all have potential for helping manage pests and minimize problems at the international level. Some pathogens such as wheat rusts, whose spores travel thousands of miles on the wind each year, require vigilance across national boundaries and sharing of information and research so that all farmers can benefit from new discoveries in research.

Environmental Impacts

Environmental implications are prominent in the discussion of numerous issues presented in prior chapters. Most often we use water and air quality as measures of environmental impact of agricultural practices. More recently the concept of soil quality and its measurement have been added to our environmental assessment, yet there are many more dimensions to a rural landscape and environment that provide the ecosystem services described above. Some of the research and education priorities in this area are given in the following sections.

At the field level, any soil loss includes loss of nutrients in the organic matter and attached to the soil particles that leave the field to concentrate in road ditches, culverts, or downstream. These must be replaced for optimum crop productivity, and there is environmental cost in the by-products of fertilizer production. Also, use of specific herbicides restricts the options for replanting of different crops or for planting noncompatible cover crops and opens the potential for drift problems during application to nearby fields.

At the farm level, there is loss of species diversity when we plant monocultures of one or two crops across the entire farm. This increases potential for pest problems and reduces the habitat for beneficial insects and pathogens that could help in biological control of unwanted species. Loss of wetlands due to installation of drain tiles brings additional acres into production but reduces the potential for biodiversity in those wetlands as well as the water storage and cleaning that goes on there.

At the landscape level, we know less about the impacts of widespread monoculture and intensive use of pesticides and soluble fertilizers. Just as on the farm level, there is loss of ecosystem services for society when the landscape is homogenized with one or a few crop species. There is also an aggregate loss of habitat area and connectivity of the remaining habitat, both crucial to maintain wildlife.

At the regional level, flooding and water quality are impacted by soil erosion and pesticide residue contamination of runoff waters and leaching. Many of the recent major floods in Iowa are said to be due to clearing of lands, draining and elimination of wetlands, and monoculture practices upstream in the regional watershed. Costs of building levees and replacing lost property have reached staggering levels in the past two decades.

At the continent level, there is silting of rivers, lakes, and dams that requires costly remediation. Research on changes in farming practices could lead to major reduction in these problems. This is coupled with disappearance of some of the best farmlands that are located near cities; these are the areas prime for development into housing, commercial use including shopping malls, and expanded transportation infrastructure.

At the world level, there is a major dead zone (more than about 20 000 km^2 [8000 sq. miles]) in the Gulf of Mexico as a direct result of contamination from the farming activities in the large Mississippi River Basin. Global warming due to emissions of CO_2 can be reduced by reductions in fossil fuel use in agriculture and other sectors of society. To change these trends much more research is needed, as well as education of the public and development of policy.

Economic Impacts

"If it's not economic, it's not sustainable"—we hear that refrain ad nauseum, but must respond to the current reality of the economic system in order to operate in the real world. Until other ecosystem services beyond food and fiber production are recognized and rewarded, it is essential that we recognize the immediate and very real constraints faced by the farmer in the design and implementation of new agroecosystems. Here are some economic priorities for research and education.

At the field level, our current commodity crops are in excess production in some parts of the world and deficient in others. For the individual farmer, it is essential to recognize the local reality of crop prices and seek alternatives that are more rewarding for the short term, and that fit into the overall farm plan. These include specialty grains, unique crops that have niche markets, and crops that can be processed or otherwise increased in value on the farm. We need to study the implications of short-term decisions that make economic sense but that contradict the quest for long-term sustainability. Even though continuous soybean [*Glycine max* (L.) Merr.] may be the most profitable crop in Nebraska, it will be impossible to produce them for more than a few seasons without degrading the soil and water resource by excessive erosion and loss of soil and fertility.

At the farm level, the diversification of enterprises can lend a degree of stability to income and reduce risk of the impacts of catastrophic failure in production or disruptions of markets for one or two major commodity crops. Alternative crops, or those that have multiple marketing opportunities, can significantly lower the risk of total failure. The economic complementarity of enterprises can also add value and stability to a farm. Growing hot peppers, onion, and garlic provides potential for direct and fresh marketing, as well as production of salsa on the farm and selling this to local or nearby markets. Growing grains and pastures, and producing a range of livestock species by stacking these enterprises gives a diverse farmscape as well as diversified income sources.

At the landscape level, there is great need to research the potentials of local markets for produce and elaborated products from the farm. In contrast to the lure of the global economy, many consumers are attracted to the concept of buying local, knowing who grows what they eat, and supporting the local economy. The proliferation of farmers' markets and grassroots food organizations certainly attests to this interest. The viability of local economies is directly tied to the success of local farms, and especially to the linkages those farms have with consumers and other local businesses. The alternative types of products and markets, as well as implications of bioregional food systems need to be studied.

At the regional level, the potentials for exchange and sales of products among communities need to be explored. There can be a relative comparative advantage of growing certain crops and animal species in specific microniches in the landscape, and this can be exploited as an integrative system linking neighboring communities, not only countries across the globe. Currently much of the poverty in the USA is concentrated in rural areas that depend on agriculture. As a result of farm consolidation and lack of critical investment in alternative businesses in many rural communities, there are people who are on government support programs or without assistance all together, and this continues to cause migration to cities. These economic patterns can be studied, and alternative solutions explored.

At the continent level, national farm support programs tend to distort the market and disfavor small and medium farms as farmers seek adequate prices and access to markets. There is currently a large and measurable impact of the NAFTA agreement on small, traditional farmers in Mexico who grow maize (*Zea mays* L.) and other crops for local consumption and sale. Their local economy has been undermined by the import of large quantities of low-cost U.S. grain, making their economic situation highly tenuous and creating situations that are not socially sustainable as well.

At the global level, there is overproduction of commodities in some areas that leads to a glut on the world market and a depression in prices worldwide. At the same time, this excess production is not reaching those without the financial resources to access this grain, or there are poor consumers who are too far from the food supply to make use of it. This distribution challenge is economic, geographical, and political, and it hampers our capacity to get food to those who need it, in spite of the reports that there is enough production today on a global basis to feed everyone.

Social Capital

Human resources on the farm, in the community, and throughout a region represent one of the most important sources of innovation in addition to the labor and management that will make systems run. Education, community infrastructure, communications, and local organizations are among the factors that represent the growth of human capital that can be used for development in the future. This social capital (see Flora, 2001a and Chapter 7, Flora, 2004) can be researched across the same levels of spatial hierarchy.

At the field level, social capital is expressed through decisions by farm managers who reflect experiences and education in farming. Appropriate, location-specific recommendations that are unique to place and condition in each field reflect the farmer's wisdom about that place. Such site-specific decisions are contrary to current conventional practices that homogenize fields with widespread, average practices and hybrids or varieties that fit the general conditions of the area. Another social dimension is the need for regulations and incentives that encourage conservation in all production fields. These function at higher levels in the hierarchy but impact practices at the field level.

At the farm level, social capital has been little researched and involves the infrastructure and potential quality of life that is necessary to provide incentives for families to live on the farm and in rural communities to sustain family farm–based agriculture. With the increasing financial pressures and stresses of moving toward larger farm sizes, there is need for study of rational alternatives that will allow a beginning farm family to get into the business as well as existing farm families to make an adequate living and meet their needs. This is a reasonable expectation for rural and city dwellers alike.

At the rural community level, the services and social contacts necessary to maintain a rich educational and healthful environment for a family are essential. There is need for research that details the factors that must be in place for such infrastructure to be maintained and for a community to thrive. Without such information it will be difficult to maintain a viable rural community to serve agriculture and the supporting activities that add value within the community and provide stable, rewarding jobs to those in the area. A loss of cultural landscape has been addressed in many of the EU countries through subsidies for reconstruction and maintenance of period farms and buildings, keeping the flavor of the landscape in vivo rather than depending on history books or museums to provide contact with history and its lessons.

At the regional level, a high degree of outmigration has depleted the rural population to the point where many schools and churches close, retail stores and farming supply firms cannot stay in business, and hospitals and other essential services move to larger towns where there is a critical mass of population. We need research at the regional level to determine how people can be adequately employed, with both economic and social motivation to stay in rural areas and enjoy an acceptable quality of life. As in much of the rest of the world, rural U.S. areas have the highest poverty levels, and we need to find alternative methods of producing a wide range of products in agriculture, supplementing this with other industries, and keeping people on the rural landscape.

At the continent level, the phenomenon of human migration has become an accelerating process that swells the populations of urban and periurban areas, causing concentration of people and human activities and all the problems associated with crowding. Now far from their food supply, people lose touch with the natural environment as well as their source of food and gradually devalue those things that are in fact essential for survival. To maintain the urban infrastructure, huge investments must be made to replace many of the services provided freely in the agroecosystem. Research that illuminates the reasons for migration and that opens viable alternatives will benefit the rural regions by slowing or reversing this migration pattern. International migrations further exacerbate problems because of costs of supporting new immigrants and costs of the loss of educated specialists and a labor force that emigrates from the exporting areas. Free movement of people and capital is a cornerstone of the emerging globalized economy, but this process in fact benefits only a minority of people so far.

At the global level, the emergence of widespread trade agreements have begun to shape the look of agriculture in a major way. Even with the hope that globalization will trickle down to the poorest of the poor and that everyone will benefit, the major winners thus far appear to be those multinational corporations that are poised and ready to capitalize on the free movement of goods and services across borders. Gross inequities in availability of food, income, and quality of life continue both within countries and especially between north and south, and the differences appear to be widening rather than lessening with the current set of agreements on trade. Research that reveals alternative and more equitable paths to development is much needed.

Ethical Issues

Ethics in agriculture were discussed by Fred Kirschenmann (Chapter 11, Kirschenmann, 2004). There are numerous issues that need to be explored, and this is one area that has been seriously neglected in both research and education in our agricultural colleges and universities. Fortunately, courses in agricultural ethics and bioethics are becoming more available in some institutions, but there are limited resources available for teaching due to limited research in the area and few faculty dedicated to these topics. Issues in ethics can be explored across the same hierarchy of spatial scale and include the following.

At the field level, the farmer or manager makes ethical decisions that influence the types of enterprises and technologies employed. Decisions on chemical pesticide or fertilizer inputs vs. organic practices may impact not only the immediate production field but also those of neighbors, if there is spray drift with the wind or runoff of chemicals that reach fields managed by others. An ethical decision is whether to produce plant or animal products, and how these will be elaborated to feed the human population. The issue of animal welfare is especially important in Europe and becoming higher priority in the USA, and humane treatment of domestic livestock is an increasingly important issue to the consumer as well. One of the first research issues addressed by scientists from Tribal colleges and universities in South Dakota was the ethical treatment of bison as a research animal. The concept of "We are all related" helped shape the research guidelines and philosophy. Studying approaches of other cultures may give insight for our own development of ethical decisions.

At the farm level, there is a vital need for more research on the intergenerational equity in land ownership and use. To maintain a viable family farm sector, it is essential to understand the ways that ownership can be passed from one generation to the next without undue financial hardship for either party. There is a need for researching the level of support needed by beginning farmers to sustain their operations as they gain equity in the land and equipment needed to sustain the farm.

How society handles these issues reflects the priorities and values that a country places on sustaining food production.

At the landscape level, there is currently a loss of cultural identity with food and natural resources, as an increasing proportion of the U.S. population lives in an urban environment. We need to understand the values that underlie decisions to live in the rural landscape and how people can generate sufficient economic activity there to sustain viable rural communities. The lack of urban–rural connections is illustrated by the difficulties that arise at the interface between farming operations and urban subdivisions. There is a lack of understanding between two contrasting groups of residents, and conflicts arise over land and resource use. Research into these issues can provide more information on how to resolve conflict and find acceptable compromise between two conflicting life styles and ways of earning a living.

At the regional level, the loss of farming families and voters causes a large shift in political power to the great disadvantage of rural residents. Lack of adequate support programs and federal farm legislation that favors only the largest of farm operations cause an acceleration of loss of family farms as those who receive payments can bid up the purchase or rental cost of land. Who should farm the land in the future is an ethical issue, and one that needs further study and viable solutions to achieve a sustainable food system.

At the continent level, current ethical statements or apparent postures by some countries do not match their administrative and legislative decisions. For example, in the USA, the "freedom to farm" legislation that was supposed to open markets and reduce farm support payments has resulted in exactly the opposite. While the USA espouses free trade as a long-term national goal, there are edicts that carefully protect specific crops and industries as farm support payments distort the world market and make if difficult to predict the future in farming.

At the global level, we continue to struggle with gross inequities in distribution of benefits in terms of both income and food availability. Globalization of trade appears to be accentuating rather than reducing these differences. Joliffe (1997) presented statistics that show the education of women plays an even greater role than food availability in preventing malnutrition in children and should make us reconsider the role of agronomy in feeding the world. Careful study of the current situation and consideration of viable alternatives must be a research priority for the near future.

The research issues described here are meant to accentuate some of those described and summarized in the preceding chapters and to expand the discussion to additional issues of importance. The role of education is to elaborate these questions and dilemmas, and to bring greater priority to the study of such issues in both the realms of research and teaching. It is through thoughtful consideration of the current trends and priorities, and study of alternate approaches used in nondominant societies, that we can rationally assess the impact of this direction on the future of the food system in the USA and globally and find alternatives that will provide a more equitable system in the future.

CONCLUSIONS

As we look for ways to make our thinking more holistic and our vision more multidimensional, we can turn to individuals and cultures that are not characterized by the linear thinking that dominates the agricultural research and education system developed in the last 120 yr in the USA. More than 50 yr ago Aldo Leopold (1949) wrote in his *Sand County Almanac*, "That land is a community is the basic concept of ecology, but that land is to be loved and respected is an extension of ethics." We can also learn from indigenous cultures that often still practice collective production, share commodities, and live in communities of extended families. In many indigenous cultures, women are the traditional food producers and nutritionists for the family. As pointed out by Ingrid Washinawatok, a young Menominee woman, after listening to a long discussion by non-Indian activists who wanted to strategize about Indian issues,

> The traditional Indian people are protecting something that is important for everyone. They are trying to keep the land alive, and the world in balance. Sometimes I get the feeling that you really don't get the point. You are not really helping us, we are helping you.
>
> (quoted in *In the Absence of the Sacred*, Mander, 1992)

A third model is found in the feminist literature that reminds us of the relationships between women and the environment. Harcourt (1997) argued that western development theory is founded on western biases and assumptions that exclude women and the environment from its understanding. A practical example of the application of this philosophy is found in the novel *Herland* by Charlotte Perkins Gillman (1979), written in the early 20th century. As an example of current priorities, a comparison of funding for national CSREES goals in 2001 shows $5.3 million dedicated to the goal of finding an agricultural system that is highly competitive in the global economy, while $1.4 million was given for greater harmony between agriculture and the environment, and $1.2 million for a healthy well-nourished population. We need to reevaluate research and extension funding strategies and priorities, giving more emphasis to those using holistic and balanced approaches to achieve sustainability. Models and examples from other cultures can help us examine creative and alternative choices for the future.

STUDY QUESTIONS

1. Discuss the political process through which research and education priorities are established, and how cultural norms and human values impact these decisions.

2. Discuss the ways in which a spatial hierarchy helps us understand the use of specific technologies to solve specific production or social challenges, and where these solutions apply. In what ways is a time scale or hierarchy useful in understanding agriculture and food systems?

3. What additional priority research and education areas would you place in Table 12–1? Conduct an exercise in which a group brainstorms this question starting with a blank table.

4. How do we arrive at consensus for future research and education priorities? How do we assure that all stakeholders have a voice in this process?

5. What are some innovative ways that nondominant and nontraditional cultures and individuals can be brought into discussion of research and education priorities for the future?

6. What specific values, practices, and systems can be learned from Native American culture and agriculture that will be useful in shaping tomorrow's agriculture in the USA? Are there unique values or characteristics related to other indigenous systems that could be useful for designing efficient and environmentally sound agroecosystems for the future?

REFERENCES

Audsley, E. (ed.) 1997. Harmonisation of environmental life cycle assessment for agriculture. Final report. European commission concerted action AIR3-CT94-2028. Silsoe Research Inst., Bedford, UK.

Barker, J.A. 1985. Discovering the future: The business of paradigms. ILI Press, Lake Elmo, MN.

Bawden, R.J. 1991. Systems thinking and practice in agriculture. J. Dairy Sci. 74:2362–2373.

Carson, R. 1962. Silent spring. Houghton-Mifflin, Boston, MA.

Cavigelli, M.A., S.R. Deming, L.K. Probyn, and R.R. Harwood (ed.) 1998. Michigan field crop ecology: Managing biological processes for productivity and environmental quality. Ext. Bull. E-2646. Michigan State Univ., East Lansing.

Checkland, P. 1981. Systems thinking, systems practice. John Wiley & Sons, New York.

Collinson, M. (ed.) 2000. A history of farming systems research. CABI Publishing, Oxon, UK.

Conway, G.R. 1985. Agroecosystems analysis. Agric. Admin. 20:31–55.

Daily, G.C., S. Alexander, P.R. Ehrlich, L. Goulder, J. Lubchenko, P.A. Matson, H.A. Mooney, S. Postel, S.H. Schneider, D. Tilman, and G.M. Woodwell. 1997. Ecosystem services: Benefits supplied to human societies by natural ecosystems. Issues in Ecology 2, Spring. ESA, Washington, DC.

Doherty, S., and T. Rydberg (ed.) 2002. Ecosystem properties and principles of living systems as foundation for sustainable agriculture—Critical reviews of environmental assessment tools, key findings and questions from a course process. Report Ekologiskt Lantbruk 32. Center for Ecological Agriculture, SLU, Uppsala, Sweden.

Flora, C.B. (ed.) 2001a. Interactions between agroecosystems and rural communities. CRC Press, Boca Raton, FL.

Flora, C.B. 2001b. The social aspects of agroecology: Agroecology and human communities. *In* Annual meetings abstracts [CD-ROM]. ASA, CSSA, SSSA, Madison, WI.

Flora, C.B. 2004. Community dynamics and social capital. p. 93–108. *In* Agroecosystems analysis. Agron. Monogr. 43. ASA, CSSA, SSSA, Madison, WI.

Francis, C. 2004. Soil dynamics, plant nutrition, and soil quality. p. 31–48. *In* Agroecosystems analysis. Agron. Monogr. 43. ASA, CSSA, SSSA, Madison, WI.

Francis, C., G. Lieblein, L. Salomonsson, J. Helenius, T.A. Breland, D. Rickerl, N. Creamer, R. Salvador, M. Wiedenhoeft, S. Simmons, P. Allen, S. Gliessman, M. Altieri, R. Harwood, C. Flora, and R. Poincelot. 2003. Agroecology: The ecology of food systems. J. Sustain. Agric. 22(3):99–119.

Francis, C., L. Salomonsson, G. Lieblein, and J. Helenius. 2004. Serving multiple needs with rural landscapes and agricultural systems. p. 147–166. *In* Agroecosystems analysis. Agron. Monogr. 43. ASA, CSSA, SSSA, Madison, WI.

Gilman, C.P. 1979. Herland: A lost feminist utopian novel. Pantheon Books, New York.

Harcourt, W. (ed.) 1997. Feminist perspectives on sustainable development. Zed Books Limited, London, and Soc. International Develop., Rome.

Harwood, R.R., A.M. Fortuna, J.E. Sanchez, J. Smeenk, and E.A. Paul. 2001. Using agroecology as a farmer's guide for cropping systems management. *In* Annual meetings abstracts [CD-ROM]. ASA, CSSA, SSSA, Madison, WI.

Joliffe, D. 1997. Whose education matters in the determination of household income: Evidence from a developing counry. Discussion Paper. International Food Policy Research Institute, Washington, DC.

Kirschenmann, F. 2004. Ecological morality: A new ethic for agriculture. p. 167–176. *In* Agroecosystems analysis. Agron. Monogr. 43. ASA, CSSA, SSSA, Madison, WI.

Kuhn, T.S. 1962. The structure of scientific revolutions. Univ. Chicago Press, Chicago.

Leopold, A. 1949. A Sand County almanac. Oxford Univ. Press, Oxford.

Mander, J. 1992. In the absence of the sacred. Sierra Club Books, San Francisco.

Odum, H.T. 1996. Environmental accounting—Energy and environmental decision making. John Wiley & Sons, New York.

Nicholls, C.I., and M.A. Altieri. 2004. Designing species-rich, pest-suppressive agroecosystems through habitat management. p. 49–62. *In* Agroecosystems analysis. Agron. Monogr. 43. ASA, CSSA, SSSA, Madison, WI.

Olson, R.K.1995. Diversity in agricultural landscapes. p. 121–160. *In* R.K. Olson et al. (ed.) Exploring the role of diversity in sustainable agriculture. ASA, CSSA, SSSA, Madison, WI.

Parker, M. 1990. Creating shared vision: The story of a pioneering approach to organizational revitalization. Dialog International, Oak Park, IL.

Rickerl, D., and C. Francis. Multidimensional thinking: A prerequisite to agroecology. p. 1–18. *In* Agroecosystems analysis. Agron. Monogr. 43. ASA, CSSA, SSSA, Madison, WI.

Rickerl, D.H., L.L. Janssen, and R. Woodland. 2001. Buffered wetlands in agricultural landscapes in the Prairie Pothole Region: Environmental, agronomic, and economic evaluations. J. Soil Water Conserv. 55:220–225.

Schlosser, E. 2001. Fast food nation: The dark side of the all-American meal. Houghton Mifflin, Boston.

Smolik, J.D., T.L. Dobbs, and D.H. Rickerl. 1994. Relative sustainability of alternative, conventional, and reduced till farming systems. J. Altern. Agric. 10:25–35.

Spedding, C.R.W. 1975. The biology of agricultural systems. Academic Press, London.

Spedding, C.R.W. 1988. An introduction to agricultural systems. Academic Press, London.

van Mansvelt, J.D., and M.J. van der Lubbe. 1999. Checklist for sustainable landscape management. Elsevier Science BV, Amsterdam.

Wackernagel, M., and W.E. Rees. 1996. Our ecological footprint: Reducing human impact on the earth. New Society Publishers, Gabriola Island, Canada.

GLOSSARY

Agricultural landscape: a landscape in which the dominant landscape element (also known as the matrix). is agricultural production (Chapter 8).[1]

Agroecology: the science of applying ecological concepts and principles to the design and management of sustainable food systems (Chapters 1 and 2).

Agroecosystem: an agricultural system understood as an ecosystem (Chapter 2).

Associated biodiversity: all soil flora and fauna, herbivores, carnivores, and decomposers that colonize the agroecosystem from surrounding environments and that will thrive in the agroecosystem depending on its management and structure (Chapter 4).

Best Management Practices (BMPs).: recommended activities carried out by those with decision-making authority (Chapter 7).

Biodiversity: all species of plants, animals, and microorganisms existing and interacting within an ecosystem and playing important ecological functions (Chapter 4).

Biogeochemical cycle: the manner in which the atoms of an element critical to life (such as C, N, or P). move from the bodies of living organisms to the physical environment and back again (Chapter 2).

Biomass: the mass of all the organic matter in a given system at a given point in time (Chapter 2).

Bioregionalism: an emphasis on the human activities and economic relationships within a defined, local area (Chapter1).

Bonding social capital: dense, internally oriented relationships oriented within the community or group (Chapter 7).

Bridging social capital: community members who reach out beyond their "comfort zone" to different kinds of people in different kinds of places (Chapter 7).

Carbon sequestration: Capturing or locking up of CO_2 from the atmosphere in sinks, such as soils (Chapter 6).

Community-level measures: values that reflect the condition of a community as a whole (Chapter 7).

Confined Animal Feeding Units (CAFO): A large number of animals are brought together in feedlots or buildings to fatten for slaughter, for breeding and birthing, or weaning (Chapter 7).

Conservation within the agricultural landscape: the planned organized management, use, protection, and maintenance of agricultural landscapes with the objective of supporting the essential physical, chemical, biological, and cultural and social functions of a sustainable agroecosystem (Chapter 8).

Consumer: an organism that ingests other organisms (or their parts or products). to obtain its food energy (Chapter 2).

Contexts: the interrelated conditions in which activities exist or occur (Chapter 7).

---

[1] Chapter numbers identify discussion of the term as defined here, although it may also appear in other chapters.

Contingent valuation method: A method for placing monetary values on nonmarket goods. This method relies on surveys of sample households in which responses are elicited about willingness to pay or to accept compensation (Chapter 6).

Corridor: an element of the landscape serving as elongated connections between patches and matrix components (Chapter 8).

Decomposer: fungal or bacterial organism that obtains its nutrients and food energy by breaking down dead organic and fecal matter and absorbing some of its nutrient contents (Chapter 2).

Desired outcome: a planned positive result of activities (Chapter 7).

Destructive biota: weeds, insect pests, and microbial pathogens, which farmers aim at reducing through cultural management (Chapter 4).

Detrivore: organism that feeds on dead organic and fecal matter (Chapter 2).

Dynamic equilibrium: a condition characterized by an overall balance in the processes of chnage in an ecosystem, made possible by the system's resiliency, and resulting in relative stability of structure and function despite constant change and small-scale disturbance (Chapter 2).

Economic sanctions: economic rewards used to coerce community members to engage in positive behavior or to desist in negative behavior (Chapter 7).

Ecosystem services: the multiple benefits of intact ecosystems that provide economic and other benefits to humans such as carbon capture, oxygen generation, filtering chemicals from water supplies, preventing floods, and others (Chapter12).

Ecosystem: a functional system of complementary relations between living organisms and their environment within a certain physical area (Chapter 2).

Emergent property: a characteristic of a system that derives from the interaction of its parts and is not observable or inherent in the parts considered separately (Chapter 2).

Energy analysis: evaluation of both the quantity and quality of energy invested in a given process in the food or industrial system, "embodied energy." (Chapter 12).

Enlightenment approach: the belief that people can be influenced to change their ethical behavior through reason and freedom (Chapter 11).

Entrepreneurial social infrastructure (ESI): when both bridging and bonding capital are high, there are high possibilities of internal collaboration, participation, and openness to constantly improve the collective condition (Chapter 7).

Environmental determinism: the theory that evolution and development of human societies and success of other species is primarily a function of geographic location and the prevalent soils and climate in that place (Chapter 3).

Environmental ethic: a code of conduct that seeks to inform behavior that will safeguard the environment (Chapter 11).

Ethical dissonance: discordant or contradictory ethical behavior (Chapter 11).

Externalities: Spillover effects from agricultural production—either positive or negative—that are uncompensated (or not accounted for in farmers' costs, if negative). in unregulated markets (Chapter 6).

Farm Bill: Legislation passed by the U.S. Congress, which sets the rules and allocates resources for food and fiber systems and rural areas (Chapter 7).

Farm plan: a plan for a farm. Can include one component (e.g., a soil conservation plan)., many components, or the whole farm. A document to allow a farmer to look forward in some way, an aid to future decision making (Chapter 5).

Farming systems analysis: looking at the farm as a functioning system. Similar to ecosystem analysis. Stocks, flows, and feedback loops can be analyzed (Chapter 5).

Financial/built capital: the physical infrastructure and monetary instruments that can be invested to create new resources (Chapter 7).

Food system: the interconnected metasystem of agroecosystems, their economic, social, cultural, and technological support systems, and systems of food distribution and consumption (Chapter 2).

Geographical information system (GIS): computer software for analyzing spatial data and presenting the results (Chapter 9).

Herbivore: an animal that feeds exclusively or mainly on plants. Herbivores convert plant biomass to animal biomass (Chapter 2).

Holistic thinking and management: conceptual or management approach that considers the whole system and the interactions among components rather than focusing on single issues and single cause–effect relationships (Chapter 1).

Human capital: the knowledge, skills, and abilities of individuals (Chapter 7).

Incentives approach: the belief that people are not likely to change their ethical behavior unless there is some personal incentive to do so (Chapter 11).

Indicators: an action or state of being that demonstrates the degree to which a particular condition exists. Indicators must be meaningful, linked to human action, relatively easily measured, and not chosen on the ease of their measurement (Chapter 7).

Initiative: procedure enabling a specified number of voters by petition to propose a law and secure its submission to the electorate or to the legislature for approval (Chapter 7).

Internalized societal norms: positive behaviors that are learned as a part of one's socialization and maturation process (Chapter 7).

Land Grant System: a system of U.S. universities established in 1862 to benefit farms, ranches, and rural communities. Traditionally black colleges and universities were added to the system in 1890, and Tribal colleges and universities were added in 1994 (Chapter 12).

Landscape: an area on the scale of several kilometers composed of interacting ecosystems (also referred to as landscape elements). that are repeated because of geology, landform, soils, climate, biota, and human influences throughout the area (Chapter 8).

Life cycle analysis: a total cost of producing and disposing of an item or process included in agricultural or industrial systems, including the infrastructure necessary for production (Chapter 12).

Matrix: the dominant element in a landscape (Chapter 8).

Monitoring tool box: a collection of tools that could be used to assess the health or quality of a farm. May include soil quality sample data, water quality, wildlife abundance, wildlife diversity. Many tools could be included (Chapter 5).

Monoculture: the cultivation or growth of a single crop or organism especially on agricultural or forest land (Chapter 7).

Multidimensional thinking: the ability to think simultaneously in several dimensions and to conceptualize challenges from multiple points of view (Chapter 1).

Multifunctionality: A perspective that explicitly recognizes agriculture has functions that are ecological and social in nature, in addition to production of food and fiber (Chapter 6).

Natural capital: soil, air, water, biodiversity, and landscape (Chapter 7).

Negative sanctions: penalizing negative behavior; used to coerce an individual to engage in positive behavior (Chapter 7).

Nutrient balance: the accounting of nutrients entering and leaving a field, farm, or other physical area and the positive or negative result of such calculations (Chapter 3).

Paradigm: a predominant set of rules, opinions, or norms that govern the thinking in a specific realm of human thoughts or activity (Chapter12).

Participatory approaches: community members identify issues, propose solutions, and evaluate actions based on an understanding of the local context (Chapter 7).

Patch: element of the landscape that is not dominant but occurs repeatedly within the matrix (Chapter 8).

Planned biodiversity: the crops and livestock purposely included in the agroecosystem by the farmer, which will vary depending on management inputs and crop spatial and temporal arrangements (Chapter 4).

Positive sanctions: rewarding an individual for positive behavior; used to coerce someone to engage in positive behavior (Chapter 7).

Primary production: the amount of light energy converted into plant biomass in a system (Chapter 2).

Principal components analysis (PCA): a diagnostic tool for detecting multicollinearity, in which there are strong linear dependencies among the independent variables (Chapter 9).

Processes: actions or operations that continue to an end (Chapter 7).

Producer: an organism that converts solar energy to biomass (Chapter 2).

Production: harvest output or yield (Chapter 2).

Productionist ethic: a code of conduct that focuses almost exclusively on producing as much as possible, regardless of the cost (environmental, social, or economic). (Chapter 11).

Productive biota: crops, trees, and animals chosen by farmers to play a determining role in the diversity and complexity of the agroecosystem (Chapter 4).

Productivity: the ecological processes and structures in an agroecosystem that enable production (Chapter 2).

Sanctions of force: certain activities, such as planning, imprisonment, or exile used to influence behaviors (Chapter 7).

Repetitive landscape unit (RLU): the minimum land area that contains landscape elements in the same proportion as the larger landscape.

Resistance: in an ecosystem, the ability to avoid or withstand disturbance (Chapter 4).

Resilience: in an ecosystem, the ability to recover following disturbance (Chapter 4).

Resource biota: organisms that contribute to productivity through pollination, biological control, and decomposition (Chapter 4).

Site-specific management: use of remote sensing GPS technologies or careful field observation to adjust input use to specific conditions in different parts of production fields (Chapter 12).

Social capital: mutual trust, norms of reciprocity, and networks (Chapter 7); human resources whose economic and other value can be enhanced through education, experience, infrastructure, and political decisions (Chapter 12).

Social control: the use of positive and negative force by humans to impose society-approved behavior on other humans in communities and agroecosystems (Chapter 7).

Social control mechanisms: the social devices used within a group to enforce social behaviors (Chapter 7).

Social system: behaviors that are deeply embedded within a community (Chapter 7).

Soil electrical conductivity: the capacity of soil to conduct or transmit electrical current. Measured in siemans per meter, and related to dissolved salts (Chapter 9).

Soil quality: the biological, chemical, and physical characteristics of a soil that allow sustainable production of plants and animals for human use (Chapter 3).

Standing crop: the total biomass of plants in an ecosystem at a specific point in time (Chapter 2).

Sustainability index: a numerical summary of factors that could be important to the long-term sustainability of the farm. Usually includes data such as resource quality (soil, water, air), wildlife habitat, human health risks, and long-term ability of farm to produce (Chapter 5).

Sustainable agroecosystems: landscapes that provide for the needs of current generations without impinging on meeting the needs of the next (Chapter 7).

Systems renewal cycle: a fundamental cycle through which all adaptive ecosystems evolve (Chapter 9).

Value chains: The links that take a good or service from producer to end user (Chapter 7).

Whole-farm analysis: similar to farming systems analysis, except that the goal is to describe more than simply ecosystem processes occurring on a farm, but also the socioeconomic forces that come in to play. Whole-farm analysis will also consider who is involved in the management decisions of the farm, economic implications of various decisions, as well as resource use, flow, and other ecosystem processes (Chapter 5).

Whole-farm planning: taking the whole-farm analysis to the next step, or planning phase. That is, based on data collected (e.g., economic, resource quantity, and quality) and the goals of the participants or managers of the farm, what are the next steps? What is the vision for this farm 5 yr from now? 20 yr from now? (Chapter 5).

SUBJECT INDEX